SUSTAINABLE
FOOD
PRODUCTION

SUSTAINABLE FOOD PRODUCTION

AN EARTH INSTITUTE SUSTAINABILITY PRIMER

SHAHID NAEEM, SUZANNE LIPTON, AND TIFF VAN HUYSEN

Columbia University Press   *New York*

Columbia University Press
Publishers Since 1893
New York   Chichester, West Sussex
cup.columbia.edu

Library of Congress Cataloging-in-Publication Data
Names: Naeem, Shahid, author. | Lipton, Suzanne, author. |
van Huysen, Tiff, author.
Title: Sustainable food production : a primer for the twenty-first century /
Shahid Naeem, Suzanne Lipton, Tiff van Huysen.
Other titles: Columbia University Earth Institute sustainability primers.
Description: New York : Columbia University Press, [2021] | Series: Columbia
University Earth Institute sustainability primers | Includes bibliographical
references and index.
Identifiers: LCCN 2021016646 (print) | LCCN 2021016647 (ebook) |
ISBN 9780231189644 (hardback) | ISBN 9780231189651 (trade paperback) |
ISBN 9780231548441 (ebook)
Subjects: LCSH: Sustainable agriculture. | Food industry and
trade—Environmental aspects. | Agriculture—Environmental aspects. |
Agriculture—Health aspects. | Sustainable development.
Classification: LCC S494.5.S86 N333 2021 (print) | LCC S494.5.S86 (ebook) |
DDC 338.1—dc23
LC record available at https://lccn.loc.gov/2021016646
LC ebook record available at https://lccn.loc.gov/2021016647

Columbia University Press books are printed on permanent
and durable acid-free paper.
Printed in the United States of America

Cover design: Julia Kushnirsky
Cover photograph: © gettyimages

CONTENTS

PREFACE

BEFORE WE START . . .

This short, introductory book, or *primer*, is about sustainability, which is simply the ability of a system to sustain its functioning indefinitely. In the case of nature, the systems of interest are ecosystems, such as grasslands, forests, coral reefs, or agroecosystems (farms). Environmental sustainability, by extension, is the ability of a system to sustain its functioning for many generations without adverse environmental consequences. A related term, *sustainable development*, concerns managing systems to improve human well-being but doing so in an environmentally sustainable way. Over the long term, ecosystems are inherently environmentally sustainable, but agroecosystems are or are not, depending on how we manage them.

The focus of this primer is on agroecosystems, that is, ecosystems that are managed primarily to produce one or more desired biological products. The dominant bioproducts of farms tend to be foods like vegetables, fruits, grains, and meat, but farms produce much more, including biofibers like cotton and silk, biofuels like dung and ethanol, and many other products of mixed utility, such as flowers, feathers, leather, coffee, tea, naturopathic medicines, tobacco, cocaine, and heroin. Irrespective of the bioproduct produced, whether maize (i.e., corn) or marijuana, an agroecosystem can be managed as an environmentally sustainable system.

Although farming has been practiced for millennia, it has not always been sustainable, and it has rarely been environmentally sustainable. Agroecosystems today dominate the biosphere, and there is a broad consensus that farming, on a global scale, is currently environmentally unsustainable. This is worrisome because billions of people are in need of improving their well-being. And given that another 2 to 3 billion people will be joining our population over the next few decades, farming desperately needs to become environmentally sustainable so that we can both improve human well-being today without jeopardizing the ability of future generations to do the same.

Before we start, we thought a preface considering the core ideas underpinning this volume and offering a guide to our primer's content and intent would be worthwhile.

THE ENVIRONMENTAL SUSTAINABILITY OF A SEALED SYSTEM

The core principles of environmental sustainability were discovered by Joseph Priestley, an eighteenth-century British theologian and natural scientist, nearly two centuries ago. Priestley discovered that if you put a mouse in a sealed jar, it eventually expires, and if you try to introduce another mouse, it dies quickly. He concluded that a mouse *injures* the air by adding something toxic to it, something he referred to as the *putrid effluvium*. What was truly astonishing, however, was that if he put a plant in the jar, after a while he could introduce another mouse and it would live. Plants somehow undid the injury to the air caused by mice. We now know that mice die in a sealed jar because carbon dioxide (CO_2) builds up and oxygen (O_2) is depleted, two processes that plants can reverse if there is sufficient light, warmth, water, and nutrients to permit photosynthesis.

Of course, Priestley never used the term *sustainability*, let alone *environmental sustainability* or *sustainable development*, but had he gone further in his research, he might well have. Imagine if he put mice and plants together in a sealed jar and then experimented to find out what it would take for such a system to persist indefinitely. In this experiment, he would have to select plants that would not only purge the system of the putrid effluvium but would also serve as food for the mice. Wastes and dead organic matter would also build up over time, so he would need to ensure that a healthy, active soil is present, one that contained the microbial species necessary to decompose and recycle these materials. If he succeeded in managing his sealed jar of animals, plants, microbes, soil, water, air, light, and temperature such that the system persisted for many generations of mice, then he would have achieved environmental sustainability.

It's possible, however, that such an environmentally sustainable jar might be a miserable place, full of scrawny mice living lives nasty, brutish, and short. If humans were in the jar, they might hope for a little more than the bare minimum. Consider the Biospherians, eight people who, in 1991, sealed themselves in a \$200,000,000, ~7 million cubic foot (~204,000 m³) controlled environmental facility called Biosphere 2. They were accompanied by hundreds of species of plants, dozens of species of vertebrates and other animals, including pollinators, and several species of domestic animals, including goats, pigs, and chickens. In many ways, Biosphere 2 was a five-star, luxury version of Priestley's jar. At the experiment's end, just two years later, the system hadn't collapsed, but invasive species and weeds had wreaked havoc, carbon dioxide levels were toxic, and many plant and animal species had gone extinct. The mental and physical health of the Biospherians declined, and they worked hard, spending half their time tending their ~0.5 acre (2,000 m²) farm to produce just enough food to eke out survival. Their management of Biosphere 2 was environmentally unsustainable, and their well-being was compromised.

Our world, the biosphere, is in some ways like Priestley's jar and Biosphere 2. It is a sealed system (in an atmospheric shell) containing many plant, animal, and microbial species—however, this system has been environmentally sustainable for billions of years, with most of its functions contributing to environmental resiliency. Currently, however, the global consensus is that our impact on the biosphere has altered it enough so that it is no longer environmentally sustainable. Invasive species are rampant, the climate is changing, mass extinction is underway, and billions of people are poor, malnourished, or unhealthy and face lives in which their well-being and those of their children and their children's children is uncertain.

We can, however, turn things around, and achieving environmental sustainability is a critical part of our endeavor to do so. This is the focus of our primer.

ENVIRONMENTALLY SUSTAINABLE FOOD AND FARMING IN FIVE CHAPTERS

There are many volumes on food, farming, and agriculture and a growing number on the sustainability of food systems. Many are guides to the practice of farming and agriculture, covering topics such as soils, irrigation, plant and animal breeding, intercropping, pest management, and more. Ours is different. This primer is on sustainable development as it relates to food, farming, and agriculture. That is, it is not an introduction to food, farming, and agriculture but a brief introduction into *how* food, farming, and agriculture fit into the framework of environmental sustainability. We view food production, farms, and agriculture as ecosystems being managed for the goods and services they provide, and we explore why they are sometimes sustainable but other times not.

It is also important to note that the social and ecological issues surrounding food, farming, agriculture, and sustainability vary in their prominence around the globe, both in terms of geography (e.g., the global South versus North America, Europe, or Asia) and socioeconomics (e.g., developed versus developing countries). The issues and examples we have selected have been drawn primarily to illustrate basic principles, but as some of the global maps provided will show, the world is very heterogeneous when it comes to food, farming, and agricultural sustainability.

The topic of sustainable food, farming, and agriculture is enormous, but we can divide it into five areas, each of which is covered in its own chapter. Each chapter is meant to stand on its own so that readers, students, and instructors can select which parts are most useful to their interests and purposes. As this is a primer, we focus on key ideas and facts and provide suggested readings for those who wish to pursue the specific topics in greater detail.

Chapter 1 considers the roots of environmental sustainability and the role of farming in human development. It begins with our origins as a species some 200,000 years ago, exploring the ecological nature of our place in the biosphere and how that changed with the advent of farming. By the end of the twentieth century, unsustainable human development was seen as the root cause of several environmental crises and of the alarming level of economic inequality around the world. In an unprecedented act of global unity, the world committed to shifting economic development to a new paradigm, one centered on sustainability.

Chapter 2 is devoted to the social and natural scientific principles of environmental sustainability. As the experiments of Priestley and the Biospherians reveal, our environment is determined by how we manage the plants, animals, and microorganisms, including ourselves and our domestic species, that govern how ecosystems function. Sustainable farming manages ecosystems to provide desired bioproducts, but it does so in a way that ensures that the many other benefits we derive from ecosystems are not jeopardized.

Chapter 3 focuses on environmentally sustainable farming in the context of the three pillars of sustainable development: environmental, social, and economic sustainability. Environmentally sustainable farming is not just about production but about improving equity and social cohesion as well as trade and policies that insure sustainability.

Chapter 4 brings in the centrality of human well-being to environmentally sustainable food and farming. Well-being isn't just about having enough calories to get through the day but about nutrition, health, and food safety; promoting sustainable livelihoods; and ensuring food sufficiency, security, and sovereignty at home and abroad.

Chapter 5 brings together the natural and social science foundations and principles of environmentally sustainable food and farming (chapters 1 and 2), their importance in sustainable development (chapter 3), and the centrality of human well-being in our deliberations (chapter 4) to produce a comprehensive framework for achieving sustainability in the Anthropocene. It reviews the policy, financial, technological, social, and behavioral issues that can either promote or hinder progress in achieving this goal.

A VIBRANT AND RESILIENT FUTURE

This primer is based on our collective experience as researchers, educators, and practitioners in environmental sustainability, in which food and farming have always played a major role. We hope our readers will find, as we have, that environmental sustainability is a fascinating, intriguing, and challenging foundation for shifting our world from its unsustainable past to a path of sustainable development. It is perhaps our greatest means for improving human well-being and the vibrancy and resiliency of our living world.

1

SUSTAINABLE DEVELOPMENT

A New Century, a New Paradigm

SUCCESS AND SUSTAINABILITY

The Momentous Occasion of Our Origin

We're newcomers to the living world. It's difficult to date just when we arrived, but at some point, a baby was born on Earth who was distinctly one of us, meaning that its genome, or its complement of genes, differed little from ours. On that day, *Homo sapiens* joined the millions of SPECIES on Earth. Our current guess is that our species was born only about 195,000 years ago. That may seem like a long time ago, but not when you consider that species in the genus *Homo* arose ~2–3 million years ago and that our apelike ancestors arose ~5–7 million years ago; apes, around 23 million years ago; mammals, around 320 million years ago; vertebrates, around 525 million years ago; and life, at least 3.5 billion years ago.

To be a species that is less than 200,000 years old is like having been born yesterday.

We were born into a global, environmentally sustainable system known now as the BIOSPHERE. The biosphere is Earth's living component, in contrast to its inanimate LITHOSPHERE (Earth's crust and upper mantle), HYDROSPHERE (all its waters, such as oceans, seas, and lakes), and ATMOSPHERE (its gaseous surface). In this context, ENVIRONMENTAL SUSTAINABILITY means functioning in a manner that does not lead to changes in Earth's physical, chemical, or biological conditions that would jeopardize life. An environmentally sustainable system isn't necessarily static, but it is resilient, meaning that it resists shocks. It is a SYSTEM in the sense that it's a complex array of interconnected entities. In this case, the entities are plants, animals, and microorganisms, and the links are the ways that each organism shares common resources like air, water, and NUTRIENTS or the way they directly influence one another through predation, disease, competition, and other BIOTIC INTERACTIONS. Over the long run, meaning centuries to millennia, the biosphere is an environmentally sustainable world.

The biosphere sounds like a big, homogeneous, planetary thing. However, it is actually quite heterogeneous, made up of distinct BIOME subsystems such as rainforests, savannas, deserts, and tundra on land and kelp forests, eel-grass beds, and abyssal plains in the oceans and seas. Biomes are composed of smaller ECOSYSTEMS, which are spatially defined collections of species that interact with one another. For example, the savanna biome refers to all grasslands around the world with well-spaced trees, and the Serengeti ecosystem refers to the savanna in Tanzania and Kenya. In the modern world, many ecosystems are *managed* ecosystems, such as urban and agroecosystems, the latter being the focus of this primer.

The Extraordinary Success of a Relative Newcomer

We may be newcomers to the biosphere, but in our short time here on Earth, we've proved ourselves to be possibly the single most successful species that has ever existed. Success, of course, is a subjective term. To be more scientific and less subjective, we'll take a quantitative approach, which means exploring the "numbers of success" in some detail.

From a biological perspective, one could argue that whichever species has the greatest mass is the most successful. Let's say the average human weighs around 137 lbs (62k). If we multiply that average mass by our current global population of 7.6 billion people, that gives us a species BIOMASS of ~471 million tons.[1] Is that evidence of success? To answer this question, let's focus on the amount of carbon (C) one finds in biomass as that's the ELEMENT that matters the most to all living things. The total mass of carbon found in all of humanity is about 60 million tons. By comparison, the total mass of carbon of all the wild mammals of the world, including elephants, hippos, rhinos, hartebeests, and wildebeests, is only ~7 million tons.[2] So, by mass, our one species has almost ten times the total mass of over ~6,000 species of mammals.

The global mass of domestic species, as opposed to wild species, however, poses a conundrum. According to the Food and Agriculture Organization of the United Nations (FAO), each year we feed, water, medicate, and care for 19.6 billion chickens, 1.4 billion cows, 980 million pigs, and a few other species of domesticated animals like ducks, turkeys, sheep, water buffalo, and goats. They provide us with 586 million tons of milk, 59 million tons of beef, 91 million tons of pork, and 124 million tons of poultry. In many rural areas in developing countries, there may be draft animals like oxen, horses, donkeys, camels, or llamas for pulling plows or transporting goods—these number ~65 million worldwide.[3] Focusing again on carbon, global livestock biomass is estimated at a staggering ~100 million tons.

One might argue that, by mass, livestock are the most successful species on Earth, but therein lies the conundrum—they exist only because we make it possible. As their success is tied to ours and ours to theirs, it might be better to think of our species as "livestock-augmented humanity," which would bring our collective mass (humans and livestock) to ~160 million tons of carbon. That's more than 20 times the mass of all other mammals on Earth.

If being the most massive population doesn't strike you as the best measure of a species' success, an alternative metric might be the proportion of Earth's resources a species has commandeered. That is, whichever species controls the most resources is the most successful. In that case, each year our species appropriates almost a third of all the vegetation that is produced on land. We have converted more than half of Earth's natural ecosystems into managed ecosystems, such as croplands, plantations, and pastures, to supply us with food, fiber, biofuels, and more. We also appropriate 74 percent of all the rainwater that falls on land for the purposes of drinking, bathing, sanitation, industry, and irrigation. And what about the oceans? Over 4.6 million fishing vessels harvest 100 million tons of seafood per year from Earth's marine habitats. It's hard to make sense of such large numbers, but if one considers that we are talking about what just *one* species commandeers, it's pretty stunning.

Let's visualize these numbers by putting on a global dinner party for 8.7 million guests (the estimated number of different species on Earth), with each guest being an ambassador for their species. Dinner is a vegan pizza containing 120 billion tons of organic carbon derived from all the new vegetation and algae produced in a year. When dinner is served, humans take a slice that is somewhere between one-quarter and two-thirds of the pizza and then pass what's left to the millions of others seated at the table, each of whom will get a tiny sliver. Then, when the water comes around, we take three-quarters of it. Not much of a dinner party, except for humanity, of course.

This extraordinary achievement, in terms of mass and the resources we have commandeered in the space of just 195,000 years, or *1/6000th of 1 percent of all the time life has been on Earth*, raises the question of how we managed to be so successful. This is especially puzzling if one considers that by other biological measures, we are a rather ordinary species. Compared to us, many species are stronger, run faster, dive deeper, reproduce faster, can eat foods that are indigestible or toxic to us, can endure conditions that we cannot tolerate, and can do things we cannot do, like fly, hear ultrasound, see into the infrared, or luminesce. What is the story of our success?

Food, Farming, and Niche Construction: The Story of Humanity's Success

There are many reasons for humanity's stunning success over so short a period. Chief among them are what natural scientists refer to as NICHE CONSTRUCTION and what social scientists call DEVELOPMENT. We'll consider both these constructs in terms of food, farming, and agriculture, which we define in box 1.1.

Niche Construction

Every species has a unique ecological NICHE, that is, a set of environmental conditions, within which it can persist. Cacti thrive in deserts, polar bears feed on marine fauna in the Arctic, and worms called vestimentiferans live on volcanic fissures in the pitch-black water miles beneath the sea.

Many species, however, do not passively occupy their niche; rather, they actively modify or construct it. Birds, for example,

BOX 1.1 FOOD, FARMING, AND AGRICULTURE

The terms *food*, *farming*, and *agriculture* are used in different ways, so for the purposes of this primer, we provide the following operational definitions.

FOOD refers to edible biological products, or BIOPRODUCTS, such as grains, vegetables, fruits, nuts, honey, mushrooms, and livestock, including aquatic animals such as fish, shrimp, and oysters.

FARMING refers to the production of desired biological products, which include food but also other bioproducts such as biofuels, drugs, spices, flowers, coffee, tea, cocoa, cotton, flax, wax, silk, leather, and feathers.

FARMS are ecosystems managed for the production of bioproducts. They vary considerably in structure, scale, and management. We often classify them into specific types, such as agroecosystems, agroforestry systems, aquaculture systems, croplands, orchards, plantations, pastures, ranches, and other systems in which bioproducts are produced.

FARMERS *and* FARM LABORERS manage farms, either terrestrial (e.g., paddies, pastures, or plantations), aquatic (e.g., aquaculture like salmon, shrimp, or oyster farming), or both (e.g., aquaponics, in which farmers grow both crops and fish in the same system).

AGRICULTURE is a much broader term that refers to the production, transportation, processing, and distribution of biological goods, as well as research, engineering, financing, regulation, marketing, and trade of biological goods. The scales of agricultural operations vary enormously, from local retailers like an independently owned grocery store, bakery, butcher, or farmers' market to global trade by Fortune 500 corporations such as Archer Daniels Midland, PepsiCo, Dow Chemical, Monsanto, and Kellogg.

build nests, bees build hives, beavers build dams, and corals build reefs. Leaf-cutter ants, ambrosia beetles, and many termite species are farmers that grow their own fungi for food.

Niche construction, or the active modification of one's niche, is the key to our success. Our ecological niche is that of an ape-like primate—a primarily tropical habitat with a diet consisting of wild herbs, seeds, fruits, nuts, mushrooms, insects, and small animals. Within such a niche, we can sustain populations that likely would only number in the tens, hundreds, or possibly thousands, depending on the habitat. Species in our genus, *Homo*, however, fashioned clothes to endure climates we could not otherwise endure, built shelters to live in habitats within which we could not otherwise survive, cooked with fire to eat foods we could not otherwise digest, and used tools to capture species we could not otherwise catch.

Human niche construction involves many efforts, from building dams and roads to digging canals and tunnels, but nothing compares to farming as a form of niche construction. Without intervention, natural ecosystems would never provide enough food for us to expand outside the small niche of a large primate. Through farming, however, we modified ecosystems to fulfill virtually no other function except supply us with the species we desire, be they edible species or species that provide us with other bioproducts (box 1.1).

Through farming, we have constructed a global niche that ranges from subzero Arctic regions to the tropics, from high elevations to coastal regions, and that sustains local populations that can number in the thousands, tens of thousands, even millions in localized communities and a global population already numbering 7.6 billion and projected to reach to 10 to 11 billion by the year 2100. As shown in figure 1.1, our constructed niche is global in extent.

FIGURE 1.1 Human success: agriculture as niche construction
on a global scale.

Humans, like all species, modify their habitats, or construct their niches,
to improve the benefits they derive from their ecosystems. Terrestrial
ecosystems, from forests (top), to grasslands and savannas (middle), to
marine systems (bottom), have been modified in many ways (moving
from left to right), but the primary change has been an increase in the
population of the species we harvest at the expense of most other species.
These changes have been instrumental to our success and are now global
in extent.

Source: Authors.

Food, Farming, Development, and Human Well-Being

SUSTAINING SUCCESS

Niche construction largely explains our success, but success is not a static outcome—it has to be sustained through constant investments. Sustaining our success, however, is a tall order. At a minimum, meaning staying just barely alive, our 7.6 billion people require 380 billion liters (80 billion gallons) of water for drinking, cooking, bathing, and sanitation as well as 3 million tons of fruits or vegetables and 15 billion calories of energy (in terms of wheat, this is 3.57 million tons) *every day*. Furthermore, for many, success is not just maintaining status quo but achieving either growth or improvement in well-being. Focusing on growth, the United Nations Department of Economic and Social Affairs estimates that our population will reach 11.2 billion people by 2100.[4] That means by 2100 we will need an additional 180 billion (4.8 billion gallons) of water, 1.4 million tons of fruits and vegetables, and 1.8 million tons of flour *every day*. The question we face today, and what motivates this primer, is how to sustain such staggering production in a manner that is environmentally sustainable and keeps development on track with respect to improving human well-being.

FARMING AS NICHE CONSTRUCTION AND DEVELOPMENT

Development, in its broadest sense, concerns the set of activities undertaken to improve our well-being. Farming, which we described earlier as a form of niche construction, is also a form of development in the sense that it is a set of activities undertaken to improve our well-being by supplying us with the bioproducts we desire. If we are simply gathering food in an unperturbed HABITAT with little in the way of long-term effects, this would be living in the absence of niche construction or development. If, however, we lay claim to land, burn it to clear away plants and animals, restock it with desired species, eliminate and keep out the hordes of plants and animals that would otherwise occupy the cleared habitat, and use the stocked animals for their milk, hides, or meat, we have engaged in human-directed development. We can go further by selecting and breeding plants and animals that are easier to manage and are more productive than their wild relatives. If what we harvest is in excess of what we need, we might hoard it or enter into trade with our neighbors. We might have to store, process, and then transport what we produce. We might grow products some desire and others abhor, like tobacco, marijuana, coca, or opium. We might employ others or force others to work for us. Conversely, we might establish guidelines, rules, or laws that protect farm laborers from exploitation or enslavement. We might drain a wetland, impound freshwater, irreversibly degrade the land, or drive endangered species extinct. On the other hand, we might regulate farming practices to prevent environmental harm. No matter how we go about procuring food, if it is more than sustainable hunting and gathering, we are engaging in niche construction (a nice, neutral, natural science term). If we call it farming, and if we consider it development, then a host of environmental, economic, and social issues emerge, which we'll consider later.

THE SHIFTING PARADIGMS OF DEVELOPMENT

Development's Old Paradigm: An Unsustainable Future

A PARADIGM is a framework of ideas that are compatible with our understanding of the way the world works. As our understanding of our world evolves, our paradigms shift. The paradigm for development before the nineteenth century and even through most of the twentieth century was based on our sense that natural resources were inexhaustible because they either vastly exceeded our demand or regenerated at rates comparable to how fast we depleted them. This development paradigm allowed for the largely unregulated exploitation of natural resources and was often accompanied by the marginalization of indigenous peoples, the exploitation of landless and impoverished peoples, and enslavement. However, with the rise of opposition to foreign exploitation of the natural resources of developing countries, widespread famine, and worrisome environmental trends such as climate change, mass extinction, and pollution, our understanding of the world and Earth's systems has evolved. Contrary to the old paradigm, natural resources are exhaustible, and where regeneration is possible, such as forest regrowth after clear-cutting or the natural recovery of a collapsed fishery, it is invariably too slow to keep up with harvests. Thus, our development paradigm has needed to shift to one centered on sustainability.

Development's New Paradigm: The Emergence of Sustainability

The emergence of sustainability as a new development paradigm is often traced to the 1987 report by the United Nations World Commission on Environment and Development entitled *Our Common Future*, led by Gro Brundtland. The Brundtland Report, as it was dubbed, painted a stark picture of our world as it neared the end of the twentieth century, but it proposed as the way out a global shift from traditional development to SUSTAINABLE DEVELOPMENT.[5] The report described the state of our planet as one of environmental decay, poverty, hardship, pollution, and diminishing resources, all trending downward to ever worsening conditions. The report linked agricultural development to desertification, deforestation, and increasing levels of toxins, especially carcinogens, in our food and water supplies and noted that although food production may have been outpacing population growth, over 510 million people remained hungry in 1985—and population growth suggested this number would only grow. Thus, while the Brundtland Report considered increases in food production as a development success, it found the environmental and social costs that accompanied this success as impetus for change. The report called for a global shift to sustainable development, especially to ensure that future generations would be able to develop as we have, a concept known as INTERGENERATIONAL EQUITY; however, the report did not provide a blueprint for how to transition to such a development path.

In 1988, just one year after the Brundtland Report was released, the FAO defined sustainable development in the context of agriculture, forestry, and fisheries (including aquaculture) as

the management and conservation of the natural resource base, and the orientation of technological and institutional change in such a manner as to ensure the attainment and continued satisfaction of human needs for present and future generations. Such sustainable development (in the agriculture, forestry and fisheries sectors) conserves land, water, plant and animal genetic resources, is environmentally nondegrading, technically appropriate, economically viable and socially acceptable.[6]

The Triple Bottom Line

While the Brundtland Report did not provide a blueprint for sustainable development, it did argue that continued success hinged on integrating across the natural, social, and economic dimensions of development.

A useful way to understand what such integration means is the idea of the TRIPLE BOTTOM LINE (TBL). This is an accounting paradigm in which the net value of an enterprise is tallied using a ledger that includes natural, economic, and social costs and benefits (figure 1.2). If the net value of an enterprise is

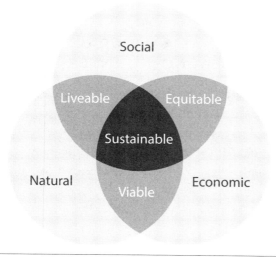

FIGURE 1.2 The four states of the Triple Bottom Line (TBL).

The TBL is an integrative framework that identifies four possible states for economic development based on the definition of sustainable development as the union of social, economic, and natural processes. Illustrated as a Venn diagram, a sustainable state exists only if all three sets of factors are integrated, creating a world that is equitable, livable (human well-being is sustained at or above minimums), and viable (environmentally sustainable).

Source: Authors.

positive across all three factors, then the enterprise is sustainable. For example, if a farm's operation generates substantial revenue from its production, whether maize, cotton, beef, or tulips, it is not deemed sustainable unless its labor practices are just and its operations do not impair natural processes or incur environmental costs.

The TBL divides the world into four states (equitable, livable, viable, and sustainable), as shown in figure 1.2. The union of any two dimensions produces worlds that may be equitable, livable, or viable, but a sustainable world occurs only at the union of all three.

FARMING AS A NATURAL, SOCIAL, AND ECONOMIC ENTERPRISE

The Brundtland Report and the TBL concept point to the importance of integrating the natural, social, and economic dimensions of development. Here, we review these dimensions individually, but we will later integrate them into a framework that illustrates the concept of sustainable farming.

The Natural Dimension of Food, Farming, and Agriculture

Taking a step back from the aspirational Brundtland Report and the abstract TBL, let's examine the fundamental science underpinning the natural dimension of sustainable development. The fundamental science is BIOGEOCHEMISTRY, which, as the name implies, concerns biologically driven geochemical processes.

We'll start first with chemistry, then consider the chemistry of biological entities, and then consider how life influences geo-chemical processes.

Biology's Chemistry

Of the ninety naturally occurring chemical elements found on Earth, life is mostly made up of the lighter elements, such as hydrogen (H), carbon (C), nitrogen (N), phosphorus (P), and sulfur (S) (figure 1.3). There are several intermediate-weight elements that are essential to many life forms, including iron (Fe), calcium (Ca), sodium (Na), potassium (K), magnesium (Mg), iodine (I), copper (Cu), zinc (Zn), and molybdenum (Mo), but life needs only tiny quantities of these elements. Heavier elements such as aluminum (Al), tin (Sn), silver (Ag), and gold (Au) are rarely found in living organisms, and others, such as lead (Pb), mercury (Hg), and uranium (U), are toxic.

Surprisingly, only six elements (C, H, N, O, P, and S) make up 90 percent of the mass of life on Earth, and only a couple of dozen are used in any appreciable amounts (figure 1.3). Consider the elemental composition of an animal, such as ourselves; a plant, such as maize; and a typical microbial organism, such as a bacterium, shown in figure 1.4.

Table 1.1 provides more detail, showing the relative abundance of elements in life, the cosmos, and our planet, divided into the litho-, hydro-, and atmosphere. What is apparent in table 1.1 is that the relative abundance of essential elements on Earth doesn't match what life needs. This imbalance between life and its surrounding environment is why farming often involves managing water and elements such as C, N, P, K, and S.

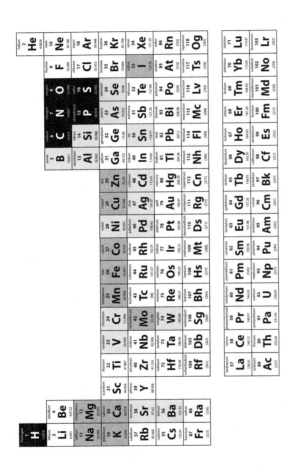

FIGURE 1.3 CHNOPS.

The periodic table of the elements, shown in this figure, organizes elements (numbered 1 through 118) according to their chemical properties and are largely ordered by their atomic mass (the numbers at bottoms of each square), which is standardly presented as the average mass of an atom of that element relative to one-twelfth the mass of an atom of ^{12}C, an isotope of carbon. Living organisms use many of the largely lighter-weight elements to make the biomolecules they need for growth and reproduction (shown here in gray or dark-gray squares). Six elements, however, are critical (black squares) to all living organisms: carbon (C), hydrogen (H), nitrogen (N), oxygen (O), phosphorus (P), and sulfur (S).

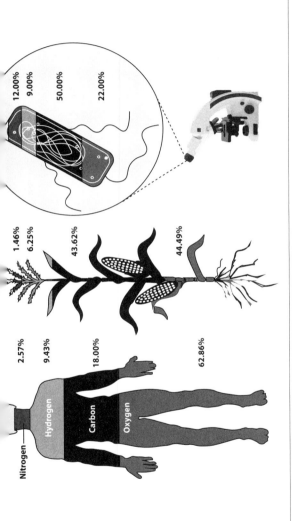

Nitrogen — 2.57%
Hydrogen 9.43%
Carbon 18.00%
Oxygen 62.86%

1.46%
6.25%
43.62%
44.49%

12.00%
9.00%
50.00%
22.00%

FIGURE 1.4 All organisms on Earth need C, H, N, O, and other elements, but how much of each element each organism needs varies considerably.

On the left, a typical person is illustrated in terms of the percentages of C, H, N, O, and other elements they are made of. In the middle, a plant, such as maize, typically shows a much larger percentage of carbon in its elemental composition. On the right, in the circle, a typical bacterium (here shown as one might see it under a powerful microscope) may be made up of a much higher percentage of carbon and nitrogen than animals or plants. In an ecosystem, every organism must acquire the essential elements it needs to grow and reproduce or it will perish. Survival, in this sense, is a matter of acquiring essential elements in the proportions one needs. In an agroecosystem, humans manage plants, animals, and microorganisms to produce food they can consume and, in so doing, increase their intake of essential elements.

Source: Authors.

TABLE 1.1 ELEMENTAL COMPOSITION OF HUMANS, MAIZE, MICROBES, THE PLANET, AND THE COSMOS

Element	Human (%)	Maize (%)	Microbe (%)	Lithosphere (%)	Atmosphere (%)	Hydrosphere (%)	Cosmos (%)
O	62.86	44.49	22.00	29.70	20.90	83.61	1.00
C	18.00	43.62	50.00	0.07	0.02	—	0.50
H	9.43	6.25	9.00	0.03	—	10.54	75.00
N	2.57	1.46	12.00	—	78.10	—	0.10
Ca	2.43	0.23	0.50	1.71	—	0.04	0.01
P	0.97	0.20	2.00	0.12	—	—	—
K	0.16	0.80	1.00	0.02	—	0.04	—
S	0.14	0.17	1.00	0.64	—	0.09	0.05
Na	0.10	0.01	0.01	0.18	—	1.05	—
Mg	0.06	0.18	0.50	15.40	—	0.13	0.06

The elemental compositions of animals, plants, and microorganisms are quite different from what is available on Earth and in the cosmos. The table compares the abundance of the top-ten life-essential elements in human biomass (excluding water) compared to what is found in maize; microbes; the three major components of our planet where life has access to elements they need for growth, development, and reproduction; and the cosmos. Note that "—" means simply that there is so little that it is effectively absent.

While plants get the elements they need in inorganic forms so long as the elements can dissolve in water and be taken up by their roots or absorbed by other plant parts such as leaves and shoots, animals get the elements they need from food, whose elements are largely in organic form. Organic molecules are made up primarily of carbon and hydrogen atoms, and biomolecules are organic molecules that also contain N, O, P, S, and other life-essential elements. Biomolecules can come in a nearly infinite variety of forms, but there are just four dominant classes of biomolecules: (1) sugars, (2) fatty acids, (3) amino acids, and (4) nucleotides. These four basic organic molecules, shown in figure 1.5, can combine to form more complex molecules. Fatty acids combine to form lipids or fats, sugars combine to form carbohydrates, amino acids combine to form proteins, and nucleotides combine to form nucleic acids such as DNA and RNA. While it is not terribly romantic to do so, food can be defined, in its most fundamental biological sense, as just packages of these four biomolecules. This definition is a bit inaccurate: minerals, VITAMINS, salts, palatability, and many other factors make the concept of food quite complex, but at base, these four biomolecules are what matter most.

In farming, managing plant nutrition to support production often means managing mineral nutrition (i.e., elements) in inorganic form (although the elements may be supplied in an organic form such as manure). However, managing livestock nutrition and ultimately meeting human nutritional needs involve managing the MACRONUTRIENT (fats, carbohydrates, proteins) and MICRONUTRIENT (vitamins and minerals) content of food, with minerals being equivalent to some of the elements (e.g., Ca, K, Mg, and P).

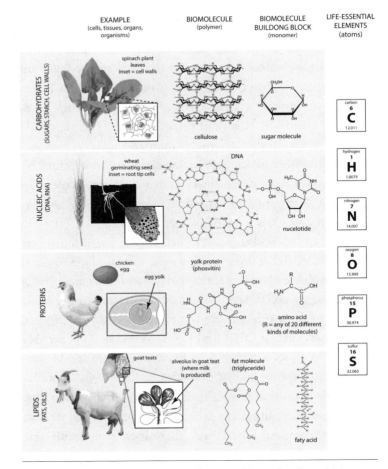

| | EXAMPLE (cells, tissues, organs, organisms) | BIOMOLECULE (polymer) | BIOMOLECULE BUILDONG BLOCK (monomer) | LIFE-ESSENTIAL ELEMENTS (atoms) |

CARBOHYDRATES (SUGARS, STARCH, CELL WALLS) — spinach plant leaves inset = cell walls — cellulose — sugar molecule

NUCLEIC ACIDS (DNA, RNA) — wheat germinating seed inset = root tip cells — DNA — nucelotide

PROTEINS — chicken egg / egg yolk — yolk protein (phosvitin) — amino acid (R = any of 20 different kinds of molecules)

LIPIDS (FATS, OILS) — goat teats / alveolus in goat teat (where milk is produced) — fat molecule (triglyceride) — faty acid

carbon 6 C 12.011
hydrogen 1 H 1.0079
nitrogen 7 N 14.007
oxygen 8 O 15.999
phosphorus 15 P 30.974
sulfur 16 S 32.065

FIGURE 1.5 Food is the source of essential biomolecules, which are sources of essential elements.

Like all animals, humans must obtain the essential elements (C, H, N, O, P, and S, shown on the right, and other elements shown in figure 1.3) needed to build the biomolecules (e.g., sugars, proteins, fats, and nucleic acids and their building blocks, shown in the center two columns) they need to grow, reproduce, and, if the supply is abundant and secure, prosper. The source of these biomolecules is food, or the plants and animals we farm (shown on the left) or harvest from the wild (i.e., fish from capture fisheries, bushmeat, wild mushrooms) and consume either directly (e.g., vegetables, meats, and dairy, raw or cooked) or in processed forms (e.g., cheese, breakfast cereals, or powdered eggs).

Source: Authors.

The Biogeochemical Cycles of Earth

There is roughly a trillion tons of carbon in the mass of life that is found throughout the biosphere. As one might imagine, if such an extraordinary mass of life is struggling to get the elements it needs to grow and reproduce, the chemical nature of the atmosphere and Earth's surface, including the oceans, are going to be strongly influenced by such a massive amount of biology. The 8.7 million species of plants, animals, and microorganisms making up this mass form immensely complex webs in which organisms feed on, compete with, facilitate, parasitize, or otherwise affect one another, but at the heart of any such web are just three classes of organisms: (1) producers, who produce biomass (organic matter) from inorganic matter; (2) consumers, who consume living biomass; and (3) decomposers, who feed on dead organic matter and convert it back to inorganic matter. Combine producers, consumers, and decomposers in an ecosystem (figure 1.6), and one has the necessary components to cycle matter between inorganic and organic forms, the most important biogeochemical cycles being those of C, N, and P.

On a global scale, these biogeochemical cycles are massive. On a yearly basis, gigatons (GT, billions of tons, or 10^{15} g) or megatons (MT, or millions of tons, or 10^{12} g) of elements cycle through Earth's spheres. Land plants, for example, move 120 GTs of C out of the atmosphere and into the biosphere every year, and respiration by plants, animals, and microorganisms moves 60 GT of C into the atmosphere every year. Looking at N, microorganisms remove 120 MTs of N from the atmosphere to the lithosphere and biosphere, 150 MT of N from the atmosphere to the hydrosphere, and 81 MT of N into the atmosphere from the PEDOSPHERE every year. Microorganisms move 32 MT of S from the hydrosphere and pedosphere to the atmosphere.

Plants on land move 74 GT of water from the pedosphere and lithosphere into the atmosphere every year through a combination of transpiration and evaporation.

These numbers may be hard to wrap one's head around, but the take-home message is that the biosphere's impact on the

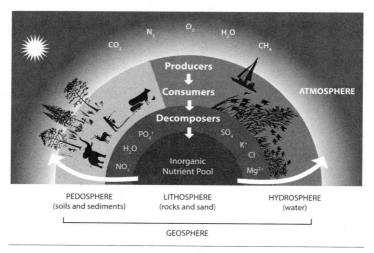

FIGURE 1.6 Biogeochemical cycles.

The presence of plants, animals, and microorganisms in terrestrial, freshwater, and marine ecosystems modifies geochemical processes that involve the movement of elements from one molecular form to another throughout the geosphere (commonly divided into the atmosphere, lithosphere, pedosphere, and hydrosphere). Such joint biological and geological processes are known as *biogeochemical processes, functions*, or *cycles*. These processes cycle elements from inorganic molecules to organic molecules and back again. Beginning with plants, or producers (at the top of the figure), inorganic forms of C, H, N, O, P, S, and other elements found in life (see figure 1.3) are assimilated and transformed into their organic forms, primarily the biomolecules that make up life. The consumption of producers by animals, and ultimately the decomposition of all organisms, leads to the transformation of organic molecules back into their inorganic forms. These biogeochemical processes move hundreds of billions of tons of elements throughout the geosphere every year.

Source: Authors.

chemistry of the lithosphere, atmosphere, and hydrosphere is immense.

With that in mind, now let's consider agriculture's role in all this. All ecosystems contribute to these immense cycles, but agroecosystems are so widespread that their cumulative impact is enormous. For example, of the 46 million square miles (12 billion ha) of land where primary producers are abundant (i.e., not ice, rock, or sand), 6 million square miles (1.59 billion ha), or 13 percent, are covered by plants we cultivate. If we consider all the producer biomass produced on Earth every year, as described earlier, humans and our livestock consume 25 percent of this. Additionally, agriculture adds 80 MT of N and 25 MT of P to farmland every year. When we look at figure 1.6 and consider our crops as producers, ourselves and our livestock as consumers, and our agricultural soils where decomposition occurs, agriculture is clearly a global biogeochemical force.[7]

The Economic Dimension of Food, Farming, and Agriculture

FIVE KINDS OF CAPITAL

Having covered the basic features of the natural dimension of sustainable development, let's examine the economic dimension.

When one owns or has assets that can be used in generating wealth, such stocks of assets are known as capital. Economists divide capital into several types, five of which are important to the concept of sustainable development (figure 1.7). (1) FINANCIAL CAPITAL consists of assets such as savings, loans, credit, or other monetary instruments that can be used in exchange, trade, or the purchase and sale of goods and services; it is perhaps the most familiar form of capital. (2) MANUFACTURED

FINANCIAL (savings, loans, credit,)

HUMAN (knowledge, know-how, skills)

SOCIAL (connections to friends, family, groups)

MANUFACTURED (machines, tools, buildings)

LABOR (physical, manual, skilled, service)

NATURAL (air, water, soil, plants, animals, microorganisms)

FIGURE 1.7 Five kinds of capital.

Assets that can be used to generate wealth are considered capital, which can be divided into different types. In sustainable development, a common typology involves classifying assets into five types of capital (*financial*, *human*, *social*, *manufactured*, and *natural*). Labor (which is often essential to manufacturing and service, though increasingly replaced by technology) is not a form of capital but is a critical factor in economic development.

Source: Authors.

CAPITAL includes machines, tools, buildings, and other assets that allow for the generation of wealth through the production of manufactured goods. Less tangible forms of capital are (3) HUMAN CAPITAL, or the knowledge, knowhow, and skills one has to produce goods or services, and (4) SOCIAL CAPITAL, such as the trust; connections through friends, families and colleagues; and membership in groups with shared values, such as religious organizations, that can impact wealth. A fifth type of capital is NATURAL CAPITAL, which we will discuss in what follows.

NATURAL CAPITAL

Natural capital is different from the other types of capital and is central to the concepts of sustainable development and environmentally sustainable farming. Natural capital includes stocks of natural assets such as air, water, soil, geologic formations, and noncultivated or nondomesticated plants, animals, and microorganisms (e.g., animal pollinators, soil decomposer microbes, animal and microbial biocontrol agents, and, on the negative side, plant pathogens, livestock diseases, and pests such as herbivorous insects and weeds). Natural capital, however, is rarely directly owned. More often, it is indirectly owned through one's landholdings. Landowners, for example, may have trees, wildlife, soil rich in microorganisms, and natural sources of water on their land, and often these assets are considered the landowners' property. Other natural assets such as air, water, migratory species, wild pollinators, and services that nature provides, such as soil formation, water filtration, greenhouse gas regulation, and nutrient cycling (discussed in chapter 2), are seldom included in the valuation of the land base (i.e., acres or hectares of property). The landowner, for example, may develop their forested landscape by harvesting and selling the trees, clearing away wild species, and becoming a farmer who grows crops to provide an annual income. While we may assume the farmer's well-being benefits from such development, without careful accounting of the natural assets the land contained before modification, it is difficult to know if the gain in financial assets is greater than the loss of the natural assets.

An alternative framework to understanding how anthropogenic activities are depleting or altering natural capital is the concept of planetary boundaries.[8] Grounded in the precautionary principle, the planetary boundaries framework provides an assessment of the risk that anthropogenic activities may perturb

Earth-system processes to the point that the Earth system changes and is no longer capable of supporting human development.[9] Nine boundaries are identified in the framework, two of which (biosphere integrity / genetic diversity and biochemical flows) have been transgressed and one of which (land-system change) is close to being transgressed. While perhaps not the sole drivers of these transgressions or trajectories toward boundary transgression, land management and agricultural practices are implicated in all three, demonstrating that unsustainable agriculture can potentially lead to irreversible Earth-system change that puts humanity and all life on Earth at risk.

Labor

The economic dimension of agriculture also includes labor, which is often where social issues of human rights and justice are important. If one has no capital (i.e., no savings, loans, skills, land, machinery, tools, or much else), one can still generate wealth by selling one's labor, usually in the form of physical or manual labor. Thus, instead of expending one's time and energy in hunting and gathering food and seeking or building shelter, one can exchange one's labor for capital from those who have accumulated it.

The Social Dimension

Finally, we examine the social dimension, which is not entirely distinct from the economic dimension, as we will see. To start, let's begin with the idea that the most important survival factor is *food sufficiency*, that is, whether there is enough food available to meet our daily needs. For the vast majority of species on Earth, not much else matters, but not so for humans. Food quality, accessibility, safety, affordability, choice, cultural suitability (some groups may have specific dietary requirements based on cultural or religious practices, such as Islamic and Jewish prohibitions against eating pork or the cultural disinclination of the majority of the Western world to eat insects), and many other factors determine what types of and how much food we consume. So too do factors concerning people's rights to land and natural resources, dietary choice, farming practices, and safeguards against labor exploitation, agricultural pollution, and environmental degradation. These and other social factors are grouped into the interrelated concepts of FOOD SECURITY, FOOD JUSTICE, and FOOD SOVEREIGNTY.

FOOD SECURITY

Food security describes the rights of all people to robust food supplies. The United Nations World Food Program (WFP) and International Food Policy Research Institute (IFPRI) identify eight factors central to food security: food should be (1) available, (2) adequately accessible, (3) reliable, (4) in sufficient supply, (5) culturally appropriate, (6) safe, (7) nutritious, and (8) capable of sustaining a healthy and active life. If any one of these factors is compromised, then food security is compromised.

FOOD JUSTICE

Food justice concerns the rights of farmers to land and the natural resources they need to pursue their livelihoods. For subsistence and smallholder farmers and for the poor who make up the majority of our global population, individual rights are integral to achieving and ensuring food security. In the Global South, especially in Latin America, where rural farmers were sometimes subjected to racism, marginalization, the dispossession of their land, and violence by large farm holders, agribusinesses, and governments, grassroots movements (e.g., la Via Campesina), rose to assert, reclaim, and protect the rights of poor farmers.[10]

FOOD SOVEREIGNTY

Food sovereignty concerns the rights of communities to choose what they eat and how they farm. To defend these rights, peasant farmers from around the world, mobilized by organizations such as la Via Campesina, le Réseau des Organisations Paysannes et de Producteurs de l'Afrique de l'Ouest, the World March of Women, World Forum of Fisher Peoples, and other organizations, met in Mali, Africa, in 2007, and signed the Declaration of Nyéléni, in which food sovereignty was defined as the right of all people to *healthy and culturally appropriate food produced through ecologically sound and sustainable methods and their right to define their own food and agriculture systems.*[11]

The Sustainable Farming Framework

The natural (or environmental), economic, and social dimensions of food, farming, and agriculture, as considered here, collectively define a framework for sustainable farming, which we illustrate in figure 1.8. Moving from left to right, we begin with the natural dimension—an ecosystem in which farmers manage plant, animal, and microbial communities to produce food or nonfood bioproducts. The farm is illustrated as an agroecosystem residing within a natural ecosystem, such as a pasture residing within a savanna, a cropland within a grassland, or an oyster farm within a coastal ecosystem. BIODIVERSITY refers to all aspects of life's diversity, be it taxonomic (e.g., number of species), genetic (e.g., number of livestock breeds), or ecological (e.g., the range of habitats a species can occupy). In the context of agriculture, AGROBIODIVERSITY consists of the array of species on and off farms or planned (e.g., crops and livestock) and unplanned (e.g., native species, exotics, or other species). On or off farms, biodiversity is responsible for biogeochemical processes and is the source of food and nonfood bioproducts and of beneficial organisms such as pollinators, BIOCONTROL AGENTS, and decomposers. It is also the source of undesired species such as pathogens and pests. The benefits humans derive from ecosystems are known as NATURE'S SERVICES, ENVIRONMENTAL SERVICES, or, simply, ECOSYSTEM SERVICES and will be taken up in the next chapter. The circle of plants and animals in the lower left represents off-farm biodiversity. Agrobiodiversity is illustrated as on-farm crop and livestock species. The rectangles represent the division of the system into its different components, with the white double-headed arrows indicating that management determines the relative extent of these divisions. We could expand an agroecosystem to the exclusion of the natural system or expand nonfood and shrink food-production areas. Human well-being, shown on the right, is determined by many factors, but food and other bioproducts are critically important. Of course, human well-being is also influenced by natural ecosystem processes, which are represented by the arrow at the bottom of the figure. The critical part of the framework is the socioeconomic system in which economics (e.g., labor, different forms of capital, trade), coupled with social factors (food security, food justice, and food sovereignty), collectively determine how farm production affects human well-being. Finally, a lot of waste is generated during the harvesting, processing, transportation, distribution, and consumption of food; thus, waste is included in the socioeconomic systems as economics, policies, technology, and social attitudes that often determine patterns of waste generation.

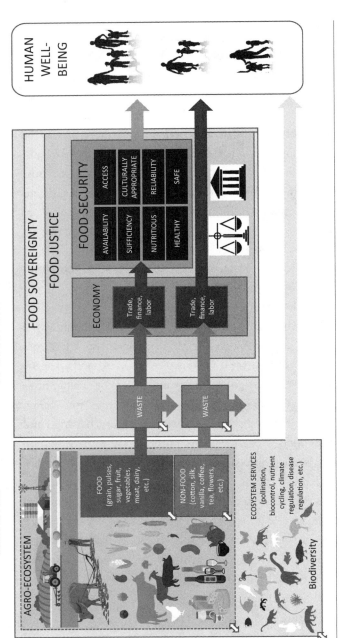

FIGURE 1.8 The sustainable farming framework.

Sustainable farming, for both food and nonfood bioproducts, requires integrating the natural, social, and economic (socioeconomic) dimensions of development. Its objective is ultimately to improve human well-being.

Source: Authors.

OUT WITH THE OLD, IN WITH THE NEW: TOWARD SUSTAINABLE FOOD, FARMING, AND AGRICULTURE

Development's Old Paradigm: Unsustainable Agriculture

THE GREEN REVOLUTION

The Green Revolution represents one of the most iconic periods of agricultural development under the old paradigm and its methods. Often described as *intense* or *industrial*, this approach to agricultural development continues to dominate large-scale farming today. The Green Revolution is a label for the late-1960s industrialization of agriculture and was initially taken up in Mexico, India, Pakistan, the Philippines, and other developing countries that had the capital (often through foreign investments) and capacity to adopt the new methodology. Green Revolution farming is characterized by using a combination of high-yielding grains, in particular wheat, maize, and rice, in conjunction with irrigation, the application of synthetic BIOCIDES (e.g., herbicides, fungicides, and insecticides) and fertilizers, and machinery to significantly increase production. Industrial farming essentially intensifies and accelerates biogeochemical processes.

The Green Revolution led to an astonishing increase in global productivity, yielding a threefold increase in food production.[12] From an environmental perspective, the intensity of Green Revolution farming meant it took less land to produce the same yields as traditional farms. An estimated 45 to 67 million acres (18 to 27 million hectares) may have been spared from agricultural transformation by industrial farming.

The Green Revolution, however, came with many environmental, economic, and social costs. Environmental costs associated with industrial agriculture include greenhouse gas emissions, contamination of waterways because of excessive use of pesticides and fertilizers, deforestation, and biodiversity loss.[13] The use of biocides, monoculture cropping systems (in which a single species is grown in a given space at a given time, even though there may be spatial and temporal variation as there is in mixed landscapes and where crop rotations are practiced), and GENETICALLY MODIFIED ORGANISMS (GMOs) creates farms and farmscapes that can be hostile environments for many species, and thus such agroecosystems may be extraordinarily low in both native species and agrobiodiversity. Rather than sparing land, the conversion to agricultural use actually skyrocketed. More land was converted to agriculture in the first thirty years of the Green Revolution than between 1700 and 1850, with almost 60 percent of arable land being converted.[14] This expansion, which continues to this day, is the primary reason for biodiversity loss.

Significant health costs were also associated with industrial agriculture. Overexposure to inputs such as fertilizers, pesticides, herbicides, and/or persistent organic pollutants has been linked to a long list of human health impacts, including diseases such as cancers, hormonal disorders, birth defects, skin disorders, and reduced cognitive abilities. Globally, the rise in obesity and diabetes has also been attributed to industrial agriculture's focus on calories rather than nutritionally balanced diets. It also narrowed the average dietary breadth of people, which lead to other nutrition-related disorders.[15]

Not surprisingly, given its mixed outcomes, the Green Revolution has been lauded and demonized. It has been seen by some as humanity's valiant effort to end hunger. It has been seen by others, however, as part of a political and economic agenda to quell class uprisings by the agrarian poor and concentrate global economic control in large agribusinesses. Whatever view one holds, the Green Revolution and the industrial farming it advanced are representative of the old development paradigm—it was not integrative and not sustainable.

Development's New Paradigm: Sustainable Agriculture

THE GREEN REVOLUTION 2.0

Industrial agriculture continues today as the dominant form of farming, though it is clearly unsustainable. In the absence of viable alternatives that are feasible under our current political and economic systems, billions of people depend on industrial agriculture. Many speak of a Green Revolution 2.0, or a Greener Revolution, as one that retains an emphasis on production and builds upon the technological advances of the Green Revolution but that more fully integrates social and economic factors into its practices.[16] Others, however, caution that retaining an emphasis on production over all other considerations will continue to distract from the importance of food security, justice, and sovereignty.

A New Management Paradigm for a New Development Paradigm

The new development paradigm of sustainability has led to a call to modify the current agriculture management paradigm to one based on ecological rather than solely technological interventions. This new agricultural paradigm is referred to as AGROECOLOGY.[17] Industrial farming is based on continuous monoculture cropping coupled with the use of synthetic

fertilizers, biocides, aggressive tilling, and often with mechanization and irrigation to maximize production. Agroecology,
in contrast, is informed by the ecology of the region, such as its
local biotic, climatic, and EDAPHIC (soil-related) conditions. Its
interventions focus on diversified cropping and either precision
application or avoidance of synthetic inputs (e.g., fertilizer
and biocides) to minimize inputs overall. Tilling is used only
when absolutely necessary, and then care is taken to minimize
erosion, such as using cover crops to protect the land between
harvest cycles.

Agroecological farming is much more aligned with the principles of sustainability than industrial agriculture. Crop diversification, for example, increases RESILIENCE to perturbations
(e.g., disease), improves pest and weed suppression, increases
pollinator availability, and improves soil health.[18] Even moving
from a monoculture to a two-species polyculture can dramatically improve sustainability. In China, for example, wheat and
fava bean INTERCROPPING (co-planting two or more species),
showed a 26 to 49 percent reduction in powdery mildew, a fungal disease that severely reduces production. Similarly, lupine
and barley intercropping in Denmark showed a 78 to 87 percent
reduction in brown leaf spot, also a fungal disease. With respect
to human well-being, crop diversification also provides for more
nutritionally balanced production.[19] Industrial agriculture
steadily increased the global availability of calorie-rich food;
however, at the same time, the global availability of macronutrients and micronutrients declined. Over the course of the Green
Revolution, for example, the protein content of global crop production declined by 4 percent, iron by 19 percent, and zinc by
5 percent. This dramatic reduction in nutrient content is attributable largely to industrial agriculture's emphasis on high-
yielding crops such as maize and wheat, which are low in

nutrient content, while the planting of nutrient-rich crops such as barley, millet, oats, rye and sorghum declined dramatically.[20]

Sustainable aquaculture and capture fisheries are similarly shifting from industrial management to management guided by ecological principles that integrate local community governance.[21]

In the next chapter, we explore these developments in greater detail.

SUMMARY

Our species is the single most successful species to have ever lived on Earth. Our success is not based on our biology, as biologically we are rather unremarkable compared to our fellow species, but because we excelled at niche construction. We clothe and shelter ourselves, eliminate predators and fight disease, and, most importantly, we farm. We manipulate plants, animals, and biogeochemical processes and entrain ecosystems to maximally produce bioproducts we desire, be they food, fibers, fuels, or flowers. Farming, like other forms of development, was based on the prevailing paradigm that natural resources were inexhaustible. But as populations grew and farming expanded and became increasingly industrial, the social and environmental costs of this paradigm began to emerge, and it fell into disfavor. By the late twentieth century, the adverse environmental and social consequences of centuries of spending down natural capital led to a call to shift development's paradigm to one centered on sustainability. Throughout the following chapters, we explore how the integration of environmental, social, and economic principles can support development, including agricultural development, that is sustainable.

2

NATURE AND NATURE'S GOODS AND SERVICES

THE SOCIOECOLOGICAL CONCEPT OF ECOSYSTEM SERVICES

As we saw in chapter 1, sustainability is a socioecological construct, and an important part of this construct (lower left of figure 1.8) is the concept of ecosystem services, also known as nature's services, environmental services, or ECOSERVICES. In economics, buyers and sellers exchange goods, which can be tangible things like food, homes, and phones or valued intangibles, such as insurance, education, and health care. Services concern this exchange of intangibles, where the sellers are service providers and the buyers are the service recipients, clients, or customers.

While we are accustomed to think of insurance companies, schools, and clinics as service providers, it's a bit strange to consider ecosystems, which are assemblages of plants, animals, and microorganisms, that way. Yet ecosystems provide some of the most important and highly valued intangibles in life, such as breathable air, potable water, fertile soils, and an equitable climate. They are also the key to sustainable food, farming, and agriculture.

In this chapter we will look first at what services are and then at ecosystems as service providers. We will use this "ecosystem as service provider" framework to explore what it takes to make unsustainable agroecosystems sustainable. This is also a great place to take a look, in broad strokes, at the state of agriculture in our world.

SERVICES

Services as Valued Intangibles

From an economic or social perspective, services are exchanges between the service provider and the service recipient. For example, an education service provides knowledge to its clients, a telecommunications company provides access to the internet, and an insurance company provides peace of mind to its policy holders. Knowledge, internet access, and peace of mind are intangible things, but we value them. Typical service enterprises include wholesale, retail, security (military, police, firefighters), transportation, storage, processing, entertainment, tourism, design, analytics, consulting, nursing, custodial, financial, legal, and government services. The distinction between goods and services, however, is not always clear, and some people lump goods and services together or refer to services as *intangible goods*. A miller who sells a bag of flour to a customer, for example, is selling one good (grain) and several services (milling, packaging, storage, and retail). To the customer, however, the exchange involves only one good—the bag of flour. Here, we will stick to the convention of considering goods as valued tangibles and services as valued intangibles, but we acknowledge that approaches differ.

The Surprising Dominance of the Service Sector

One might think that producers, who are at the foundation of most economies, would comprise the dominant economic sector, but in actuality, it's the service sector that dominates. In 2016, an estimated 80.3 percent of the U.S. workforce was employed in the service sector, such as wholesale, retail, education, hospitality, transportation, and government. In contrast, only 12.6 percent of the workforce was in mining, manufacturing, and construction—businesses that actually produce things. Surprisingly, only 1.5 percent of the U.S. workforce was employed in agriculture, forestry, and fishing.[1] The same is true worldwide. In 2017, the UN Conference on Trade and Development found that three-quarters of the workforce in developed countries and over half in developing countries are in the service sector.[2]

Nature's Goods and Services

Given their inherent value to everyone everywhere, nature's services should dominate our economies, even more than the service sector does. But they don't. In fact, they are invisible in world economics, which is the equivalent of treating them as free. In this chapter, we explore nature's services in greater detail. We will discuss how they relate to agroecosystems and explain why sustainability is absolutely dependent on making ecosystem services visible to the public and to policy and decision makers. We close with some environmental- and ecological-economic analyses that illustrate what happens to our economies when the ecosystem service sector is included in our economic calculations.

ECOSYSTEM SERVICES

Ecosystem goods and services concern the characteristics of the ecosystems we value. Recall that an ecosystem consists of a spatially defined assemblage of plant, animal, and microbial species whose collective biological processes determine the characteristics of an ecosystem. Ecosystems vary in their characteristics, which include composition, structure, functions and processes, services, disservices, and dynamic properties. Box 2.1 summarizes these characteristics.

Every ecosystem exhibits a unique set of characteristics, and there are hundreds of different kinds of terrestrial, freshwater, and marine ecosystems. At the most basic level, there are just two types of ecosystems, terrestrial and aquatic. At global scales, these two classes are divided into biomes. Terrestrial biomes are defined by climate and vegetation (e.g., tropical rainforest, desert, tundra), while marine systems are defined by location and depth, such as coastal, estuarine, open ocean, or abyssal plain. Biomes can be further divided into more specific ecosystem types. Forests, for example, can be rain, temperate, or boreal forests. Each of these can be divided further into even more specific types, such as temperate-rain, temperate-deciduous, and temperate-coniferous forests. Ecosystem classifications can be quite involved, sometimes very specifically into ECORE-GIONS. The U.S. Environmental Protection Agency (EPA), for example, identifies 967 ecoregions in the United States; the World Wildlife Fund (WWF) uses a different methodology to identify 867 ecoregions.[3]

As each ecosystem, scaled from biome to ecoregion, exhibits a unique set of characteristics, they each have a unique *service portfolio*, or unique set of valued goods and services. If the system's characteristics change, it follows that its service portfolio will change. Many factors can change an ecosystem's

BOX 2.1 ECOSYSTEM CHARACTERISTICS

Ecosystems, whether managed, like agroecosystems, or relatively unmanaged, like wildlife reserves, vary in their characteristics, which are described here starting with the term *ecosystem* itself.

Ecosystem: a spatially defined collection of plants, animals, and microorganisms that interact with one another. An ecosystem can range from highly managed (e.g., industrial farms) to less managed (e.g., a tropical rainforest reserve).

ECOSYSTEM COMPOSITION: the assemblage of species that occupy an ecosystem.

Ecosystem structure: the pattern of interactions (e.g., predation, competition, facilitation) among organisms in an ecosystem and the VARIABILITY of that pattern over space and time. For example, an ecosystem with few species that are homogeneously distributed over space and time and interact little with one another is simple in structure (e.g., an industrial farm). Ecosystems with complex structures, such as a tropical forest reserve, are the converse.

Ecosystem functions and processes: biologically driven chemical and physical processes that result in the flow of materials and energy through and the cycling of nutrients within an ecosystem. The magnitude and stability of ecosystem functions and processes are governed by the composition and structure of the ecosystem.

Ecosystem services (or environmental services or eco-services) are the functions, processes, and properties of ecosystems that positively influence human well-being.

ECOSYSTEM DISSERVICES are the functions, processes, and properties of ecosystems that negatively influence human well-being.

ECOSYSTEM DYNAMIC PROPERTIES: the dynamic properties of an ecosystem's structure and function, such as its resilience (ability to recover from an environmental shock), resistance (how much change it exhibits when shocked), and variability (how predictable its structure, functions, and processes are across space and through time). Other properties include the ability to resist invasion by exotic species and the ability to reduce the spread of pests and disease.

characteristics. Natural changes in climate, hydrology, and fire frequency, for example, can cause a rainforest to become a savanna, dramatically altering its characteristics and service portfolio. People can also change an ecosystem's characteristics through management, whether through the extirpation of predators, the hunting of game, fishing, fence construction, dam building, the digging of irrigation canals, burning of plant cover, fertilizing, tilling the soil, or the use of biocides. Agriculture is simply a natural ecosystem transformed through management rather than by natural causes.

Ecosystems vary enormously in their characteristics and service portfolios. Forests provide timber; open oceans are a source of fish; wetlands degrade organic contaminants in freshwater, regulate climate, and prevent soil erosion; and virtually all ecosystems produce oxygen, cycle nutrients, and are valued for their cultural significance. Hundreds of benefits have been identified and described, and, just as with ecosystems, many ecosystem service typologies have been developed. For example, ecosystem service typologies have been developed by The Economics of

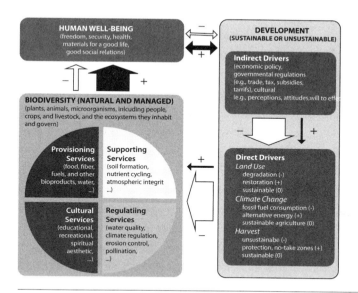

FIGURE 2.1 The Millennium Ecosystem Assessment (MEA) framework.

In the MEA framework, ecosystem services are provided by biodiversity, which directly influences human well-being (HWB).

Source: Authors.

Ecosystems and Biodiversity (TEEB), the United Kingdom National Ecosystem Assessment, and the Common International Classification of Ecosystem Services (CICES).[4] Here we have used the four-category scheme (provisioning, regulating, supporting, and cultural services) used by the Millennium Ecosystem Assessment (MEA),[5] one of the most widely used ecosystem service typologies (figure 2.1).

PROVISIONING SERVICES

PROVISIONING SERVICES are the goods that ecosystems provide to humans. Some people lump goods and services together and just refer to services. The MA took this approach. These services include food and fresh water as well as goods such as fiber (e.g., bamboo, cotton, hemp, silk), wood (e.g., timber, fuelwood), energy sources (e.g., fuelwood, but also goods such as dung and biofuels), and medicine (e.g., medicinal plants or raw materials for pharmaceuticals). Genetic resources (e.g., genetic diversity) and ornamentals (e.g., flowers, plants) are also considered provisioning services.

REGULATING SERVICES

REGULATING SERVICES are ecosystem characteristics that reduce variability or otherwise promote system stability. A prime example is the ecosystem regulation of Earth's climate system through biotic controls such as nutrient cycling, surface reflectance (i.e., albedo), and GREENHOUSE GAS (GHG) regulation. Other regulating services include the regulation of hydrological regimes, water purification, erosion control, regulation of plant growth, fruit and seed production via pollination, and regulation of disease, pests, and pathogens via biological control.

CULTURAL SERVICES

CULTURAL SERVICES are valued intangibles, but unlike other services, they are defined by peoples' perceptions and beliefs, which may change as cultures evolve. Ecosystems and the organisms within them can provide spiritual, religious, aesthetic, educational, and recreational benefits. These cultural values are important to well-being, to physical and mental health, and to inspiration and creativity. Cultural services also include traditional knowledge (or indigenous or local knowledge), which is community-based knowledge passed from generation to generation. Such knowledge incorporates cultural expressions (e.g., music, dance ceremonies, symbols) as well as knowledge of local environments (e.g., ethnobotanical, climate) and subsistence practices and technologies relevant to agriculture or hunting and gathering. Finally, an important cultural value of ecosystems is the sense of place that an ecosystem provides—a sense of belonging to a particular habitat or landscape.

SUPPORTING SERVICES

SUPPORTING SERVICES indirectly benefit humans by making other goods and services possible. They are the foundational processes underpinning ecosystem functions. Examples include soil formation, nutrient cycling, primary production, production of atmospheric oxygen (O_2), providing habitat for biodiversity, and more.[6]

Conversions of Natural to Agroecosystems

Using the MA ecosystem service framework, we explore five broad classes of natural-to-agroecosystem conversions that cover most food production systems. Figure 2.2 illustrates the extent of the transformations, providing a sense of how much we may have gained and lost in the way of ecosystem services. In each case that follows, we consider the natural system's characteristics, its service portfolio, and how these change under management.

FORESTS PLANTATIONS AND ORCHARDS

Forests Forests are tree-dominated systems that cover nearly a third of Earth's terrestrial surface. Over half of the world's forests are in tropical (42 percent) and subtropical (8 percent) regions, with the rest in temperate (26 percent) and boreal (22 percent) regions.[7] They are integral to the livelihoods of nearly

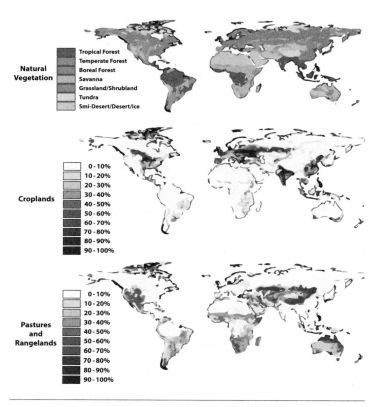

Natural Vegetation
- Tropical Forest
- Temperate Forest
- Boreal Forest
- Savanna
- Grassland/Shrubland
- Tundra
- Smi-Desert/Desert/Ice

Croplands
- 0 - 10%
- 10 - 20%
- 20 - 30%
- 30 - 40%
- 40 - 50%
- 50 - 60%
- 60 - 70%
- 70 - 80%
- 80 - 90%
- 90 - 100%

Pastures and Rangelands
- 0 - 10%
- 10 - 20%
- 20 - 30%
- 30 - 40%
- 40 - 50%
- 50 - 60%
- 60 - 70%
- 70 - 80%
- 80 - 90%
- 90 - 100%

FIGURE 2.2 Ecosystem transformations: from natural to agricultural.

All agroecosystems are descendants of natural systems whose original ecosystem characteristics have been changed by management. The primary emphasis in agricultural transformation is the enhancement of provisioning services, usually just the production of one or a few desired bioproducts such as crops or livestock. Sustainable systems retain their full portfolio of services.

Source: Based on an analysis by J. A. Foley, R. DeFries, G. P. Asner, et al., "Global Consequences of Land Use," *Science* 309 (2005): 570–74.

a quarter of the world's population. The provisioning services of forests include food (e.g., fruit, vegetables, mushrooms, bush-meat, freshwater fish); fresh water; fiber (e.g., paper pulp); raw materials (e.g., timber, palm leaves); energy resources (e.g., fuel-wood); and genetic, medicinal, and ornamental resources. They are highly diverse, being home to 75 to 80 percent of terrestrial biodiversity.

As major sources, sinks, and storehouses of C, forest regulatory services include climate regulation. They also regulate watershed outflow, both in volume and quality; influence local precipitation and hydrological cycles; prevent erosion on slopes; and serve as sources of pollinators.

Forest supporting services include soil formation (vegetation and animals contribute to biological, chemical, and physical soil weathering processes), nutrient cycling (e.g., N, P, C), primary production, production of O_2 (tropical rainforests are estimated to produce 20 to 30 percent of O_2 globally), and habitat provision for terrestrial and freshwater organisms.

Forest cultural services include recreation and aesthetic benefits and are often considered places to honor one's heritage and traditions (e.g., ceremonies, graveyards and other sacred places, sharing traditional knowledge).

Plantations and orchards Forests are managed in a variety of ways, from clear-cutting and selective harvesting to conversion to pastures, croplands, plantations, and orchards. Here, we focus on plantations and orchards, or *planted forests*, as such managed ecosystems continue to be dominated by woody plants. The FAO estimates that planted forests account for 8 percent of global forest cover, or 2.5 percent of total global land cover.[8]

Planted forests are managed to enhance the provisioning of desired bioproducts, including fruit (e.g., apricots, avocados, bananas, cherries, citrus fruits, coconut, nectarines, olives, peaches, plums, pomegranates), false or accessory fruits (e.g., apples, figs, pears), nuts (e.g., acorns, chestnuts, hazelnuts), sap (e.g., used in syrups or latex), seeds (e.g., almonds, cacao, cashews, coffee, pecans, walnuts), spices (e.g., allspice, cinnamon, cloves, nutmeg), and tea. Processed products include butters (e.g., shea, cacao, coconut, nut and seed butters), chocolate, flours (e.g., coconut, nut and seed), jams and jellies, nut and seed milks, oils (e.g., avocado, coconut, olive, palm), syrups, and vinegars (e.g., apple cider, coconut, plum). Like their predecessors, they also provide timber, fuelwood, biofuels (e.g., biodiesel from palm oil), paper pulp, medicines, and ornamentals.

Regulating, cultural, and supporting services are generally severely reduced from what the original forests provided. They can still, when managed to do so, harbor genetic resources, serve as watersheds that can regulate the flow and quality of fresh water, and minimize erosion by stabilizing soils. They also can contribute to climate regulation, but not if they are frequently harvested or burned. The cultural values of the original forests are usually lost as indigenous forest tribes and their traditional knowledge disappear, except where cultures have subsequently incorporated plantations and orchards as valued livelihoods and lifestyles.

RANGELANDS PASTURES AND LIVESTOCK SYSTEMS

Rangelands Terrestrial systems that lack or have too few trees to be called forests are often classed as RANGELAND. Rangeland is a broad category of land cover that includes deserts, grasslands, savannas (most often a mix of grasses and trees), shrublands, tundra, and wetlands. Rangelands cover a significant portion of terrestrial Earth: estimates range widely, from 30 to 80 percent.

Rangelands are often home to a diverse array of grazers, including large familiar species like antelopes, cape buffalo, zebra, wild asses, elephants, rhinos, hippos, oryx, and ostrich in Africa; ibex, saiga, yak, wild boar, and Przewalski's horse in Asia; bison, pronghorn, and antelopes in North America; llamas, vicuña, and capybaras in South America; kangaroos in Australia; and caribou, elk, and reindeer in Arctic and sub-Arctic habitats. The many small grazer species include prairie dogs, rabbits, hare, squirrels, zokor, marmots, and chinchillas, and insect herbivores like grasshoppers and locusts can sometimes rival vertebrates in terms of their consumption. Well-known predatory species, like bears, large cats (lions, leopards, cheetahs, jaguars, cougars), and canids (wolves, foxes, coyotes, wild dogs), and smaller predators, like badgers, weasels, snakes, owls, hawks, and others, regulate herbivore species' abundances in rangelands.

Rangeland provisioning services include being sources of game, pollinators, biocontrol, medicinal plants, and ornamentals, and they are repositories of genetic diversity for many domestic crop species whose ancestors reside in these habitats.

Like forests, rangeland regulates climate by sequestering C and storing it in roots and soil. Both rangeland vegetation and soils regulate water flow and purity and stabilize soils, thus reducing erosion from wind and rain.

Rangeland supporting services include soil formation, nutrient cycling, primary production, oxygen production, and habitat for biodiversity.

Culturally, rangelands provide recreational opportunities, especially ecotourism and trophy hunting, as well as aesthetic benefits and educational opportunities, and they are sites of cultural heritage for indigenous people worldwide.

Pastures and other livestock systems Rangelands can be converted to a number of different managed ecosystems, but here we focus on pastures and other livestock systems. Pastures are a dominant agroecosystem, covering 26 percent of Earth's land surface. In contrast to rangelands, vegetation in pastures is specifically managed to maximize the growth of mostly cattle, sheep, and goats. Pigs, hogs, chickens, turkeys, rabbits, geese, and ducks are also raised in rangelands, though generally in pens, barns, coops, or other protective enclosures. These systems provide food (e.g., meat, offal, bones, blood, gelatin, milk, eggs); fiber (e.g., wool, alpaca fleece); animal hides, skins, and plumage; energy sources (e.g., dung, manure); and nutrients to fertilize plants (e.g., N, P, and K from manure and urine). The animals are primarily raised for food, but they also represent genetic resources (e.g., breeds) and medicine (e.g., model organisms for the study of disease, production of vaccines, and pharmaceutical ingredients).

The portfolio of services is strongly dependent on management practices. These include the extent to which tilling, mowing, fertilizing, irrigating, seeding with non-native species, and weed control are used; animal stocking densities; feed sources; finishing systems (e.g., grass versus feedlot for cattle); pasture productivity; animal productivity; grazing methods (e.g., continuous stocking, rotational stocking, deferred grazing, seasonal grazing); grazing periods; and rest periods.

In the developed world and increasingly in the developing world, Animal Feeding Operations (AFOs) are intense forms of management that provide virtually nothing but provisioning services. In these systems, livestock, feed production, and waste management are separated and concentrated in confined areas. For example, CONCENTRATED ANIMAL FEEDING OPERATIONS, or CAFOs, typically consist of 1,000 animal units (1 unit = 1,000 pounds live weight of an animal) confined to limited spaces. Typical rangeland cattle density is 0.5 per acre (0.2 per ha), whereas a CAFO can have over 160 per acre (300 per ha) for over

45 days. Livestock in animal feeding operations are generally fed grain and silage (stored, nondried grasses). The confined quarters promote infectious disease transmission, which necessitates the prophylactic (preventative) administration of antibiotics. Waste from these operations is collected and stored in manure lagoons, which is later transported and used as field fertilizer, but when poorly managed, these lagoons release greenhouse gasses, contaminate groundwater, and overflow contaminating neighboring habitats.[9]

Clearly, the extent to which pastures and livestock systems provide regulating and supporting services is entirely dependent on management, with low-density pastoral systems providing a more balanced portfolio of services and industrial AFOs providing virtually nothing but provisioning services and net deficits in regulating, supporting, and cultural services.

Cultural services derived from pastures and livestock systems vary enormously. In low-density systems, they can include recreational, spiritual, religious, aesthetic, and educational benefits derived from cultures that have made their livelihoods dependent on livestock. The animals themselves may directly provide such benefits as sacrificial objects or as objects of worship; they may also be treated as subjects with whom humans can develop emotional attachments. Pastoralism predates Christianity, Islam, and Judaism and influenced the development of these religions in the Levant and Arabian Peninsula. Herders and pastoralists often develop a strong sense of place as they move with their animals across the landscape. Examples include the vertical transhumance systems in the Alps and Himalayas; sheep herders of the Mongolian steppes; Sámi reindeer herders of Scandinavia and Russia; Maasai pastoralists of East Africa; and the gaucho, llanero, cowboy, and other horse-based cultures of the Americas. It's difficult to know whether there are cultural values associated with AFOs, but our guess is that there are not.

FORESTS AND RANGELANDS TO CROPLANDS

Both forests and rangelands can be and have been converted to croplands, a dramatic transformation: croplands have neither the woody species of forests nor the herbivore-predator food webs of rangelands.

Croplands Estimates by the UNITED NATIONS CONVENTION TO COMBAT DESERTIFICATION (UNCCD) indicate that croplands represent only 14 percent of global land cover,[10] with cereal crops (e.g., barley, maize, millet, oats, rice, rye, sorghum, wheat) constituting both the majority of production and majority of human dietary energy supply (i.e., they serve as staple crops).

Many croplands are monocultures or single-species crops. These cater to industrial farming because they allow for mechanization, easier application of inputs (e.g., fertilizers and biocides), and achieve economies of scale to optimize return on investment and open access to large markets. These management practices are widespread. In 2007, U.S. farming included 408 million acres (165 million hectares) of land for monoculture crop production; approximately one-fifth of the United States' land area. Corn monoculture totaled 84 million acres, 73.8 million acres were in soybeans, 55.7 million acres were hay, 45.7 million acres were wheat, 9.5 million acres were cotton, 3.9 million acres were sorghum, and 2.6 million acres were rice.[11] Humans did not directly consume the majority of these crops, but instead over 70 percent of the total harvest, corn and soybeans in particular, was used as feed for livestock, fish, and poultry; to produce oils and syrups; and as biofuel. Corn and soybeans are also the top users of pesticides in the United States, and corn is the top user of nitrogen, potassium, and phosphate fertilizers. Soybean production is also popular in South America, with Brazil devoting 89 million acres (36 million hectares) of its land and Argentina devoting over 42 million acres (17 million hectares) of its land to that crop in 2018.[12]

Other important crop species include roots and tubers (e.g., cassava, taro, and yams); oilseed crops (e.g., rapeseed, sunflower); and PULSES such as dried beans, lentils, and peas (e.g., black-eyed peas, chickpeas/garbanzo beans, kidney beans). In addition to staple crops, croplands provide fruits, vegetables, other types of cereals or PSEUDOCEREALS (e.g., amaranth, buckwheat, spelt, teff), other types of oilseeds (e.g., flax, hemp, safflower), flowers used for culinary purposes (e.g., hibiscus, hops, lavender, rose) and herbs and spices (e.g., basil, ginger, mint, saffron, turmeric, vanilla). Other provisioning benefits from croplands include fiber (e.g., cotton, flax [for linen], hemp), medicinal plants and pharmaceutical ingredients (e.g., *Echinacea* species, ginseng, milk thistle), energy sources (e.g., biofuels), genetic resources (e.g., agrobiodiversity, propagation of heirloom fruits and vegetables), and ornamentals (e.g., larkspur, roses, lilies, sunflowers, tulips, zinnias).

Regulating services associated with climate and water are typically offset by the negative impacts cropland systems have on the climate system (via the emission of GHGs) and water resources (e.g., depletion of groundwater resources). The potential to store carbon in these systems is reduced because plants are harvested at regular intervals. However, croplands do have the capacity to store carbon in soils if appropriate soil management and conservation practices are used.

Supporting services are typically few, but, depending on scale and management, they can include soil formation, nutrient cycling, and the production of oxygen.

Cultural services can occur where croplands provide spiritual benefits, provide a sense of place, or have aesthetic appeal. The cultural values of the original systems are often lost, and it is unlikely that industrial monocultures provide any.

DRYLANDS AND IRRIGATED DRYLANDS

We often forget that much of terrestrial Earth is arid. Drylands, arid systems in which precipitation is generally equal to or lower than evapotranspiration (the combined evaporation and respiratory water loss of plants), cover nearly 40 percent of terrestrial Earth and are home to over 2 billion people. Though they may contain localized regions where water is available, such as oases, wetlands, ponds, rivers, and streams, these areas represent small fractions of the landscape.

Provisioning services include goods derived from wildlife such as meat, milk, and animal hides; limited supplies of timber and fuelwood; and medicinal plants. A number of the world's biodiversity hotspots (i.e., containing disproportionate numbers of species) are in dryland regions, such as the Horn of Africa and the Succulent Karoo in South Africa and Namibia. Many drylands have ecologically complex, thin yet biologically active soil crusts containing cyanobacteria, lichens, mosses, fungi, and bacteria. These crusts minimize erosion, capture water, and can fix both C and N, thereby contributing to ecosystem productivity and soil fertility. In spite of being sparsely vegetated, their soils store as much as 27 percent of the global soil C pool.

Supporting services, however, are limited given their lack of water.

Drylands provide recreational and educational opportunities. They also provide spiritual and religious benefits and support traditional knowledge systems. In some dryland systems, complex socioecological systems have been in place for thousands of years (e.g., transhumance and nomadic pastoralism), producing rich sources of traditional knowledge and community-based approaches to interacting with the environment.

Irrigated drylands Since the 1960s, irrigation has played an increasing role in dryland productivity. Irrigated drylands have doubled since the 1960s and now represent 20 percent of the land under cultivation.

Irrigated drylands are essentially croplands, providing food, fiber, energy sources, medicinal plants, genetic resources, and ornamentals. Food crops cultivated on irrigated drylands include cereals, legumes, fruits, and oilseeds. Irrigated dryland agriculture, however, draws heavily on water resources and thus disrupts hydrologic services by depleting water resources and altering natural flows. With respect to climate regulation, irrigated drylands, like other managed systems, may serve as either sinks or sources of C depending on management practices. For example, where irrigation is used, there may be a net increase in organic C inputs to soils because of the increase in biomass and introduction of C into soil. However, soil C is lost if irrigation yields significantly wetter soils because decomposition and soil respiration increase.

COASTAL SYSTEMS AND OCEANS TO FISHERIES AND AQUACULTURE

Coastal systems Coastal systems constitute vast regions, but quantifying them with a "percent cover" amount is difficult. Globally there are about 620,000 km (372,000 miles) of coastline, with 2.4 billion people (32 percent of the world's population) living within 100 km (60 miles) of a coast. Urbanization in coastal areas continues to increase, with many of the world's megacities located in them. Examples of coastal ecosystems include coral reefs, mangroves, estuaries, seagrass beds, kelp forests, deltas, salt marshes, mud flats, rocky intertidal zones, beaches, sand dunes, and coastal heathlands.

Coastal systems provide a variety of goods, including food (e.g., fish, shellfish, and seaweed), fiber (e.g., reeds), raw materials (e.g., timber), medicine and pharmaceutical ingredients, and ornamental objects (e.g., seashells, pearls). Coastal systems contribute to climate regulation through CARBON SEQUESTRATION. Termed BLUE CARBON, coastal systems such as mangroves, tidal marshes, and seagrass beds store large amounts of C below ground in soils. A smaller amount of C is stored in the aboveground biomass of these systems as well. The estimated C sequestration per hectare is greater than that of forests, though forests occupy a much larger percentage of Earth's surface cover. Coastal systems also regulate extreme events by attenuating the effects of hurricanes/typhoons, tropical storms, and tsunamis.

Culturally, coastal systems provide a range of services that include recreational, spiritual, aesthetic, educational, and social benefits. Humans have a long history of using and/or inhabiting these systems (e.g., Polynesian, Pacific Northwest Coast indigenous peoples, the Hebrides). Recreational services range widely, from swimming and surfing to whale watching and fishing. Supporting services provided by coastal systems include nutrient cycling, primary production, O_2 production, and habitat provision for both terrestrial and marine organisms.

Oceans Oceans cover 70 percent of Earth's surface, contain 97 percent of Earth's water, and constitute 99 percent of the habitable space on Earth. Oceans provide an array of provisioning services, such as food (e.g., fish), medicinal and pharmaceutical ingredients used in traditional Asian healing modalities and more recent Western medicine drug development, energy sources (e.g., offshore wind, oil, gas, thermal), and biomaterials (e.g., biopolymers used for medical purposes). While there are fewer marine species than in terrestrial systems, oceans teem with life and are a rich source of genetic diversity, with higher phylogenetic diversity than terrestrial systems. In addition, oceans regulate climate by absorbing both heat and water. Oceans, since the 1970s, have absorbed 93 percent of the excess heat trapped at Earth's surface caused by increasing GHG emissions. Oceans also are estimated to have absorbed 40 percent of anthropogenic CO_2 emissions since the start of the Industrial Revolution (c. 1800). Oceans regulate climate and hydrologic regimes (e.g., monsoons) through coupled ocean-atmosphere dynamics associated with ocean surface temperatures such as evaporation, convection, and cloud formation.

Ocean-derived supporting services include nutrient cycling (e.g., N, P, Fe), primary production by marine phytoplankton, production of O_2 (estimated to contribute 50 percent of O_2 production globally), and provision of habitat for marine organisms.

Oceans are a part of many people's cultural heritage and support social networks via shared recreational opportunities and livelihood strategies (e.g., fisherfolk).

Fisheries and aquaculture As in terrestrial systems, fisheries and seafood farming, or AQUACULTURE, mostly provide provisioning services, with other types of ecosystem service delivery being strongly dependent on management. Fisheries and aquaculture systems occur in both marine (i.e., coastal and ocean) and freshwater (e.g., lakes and rivers) environments.

Fisheries (also referred to as "capture fisheries") are systems in which fish are harvested for commercial or recreational purposes (the fish may be raised or wild caught). They are classified as industrial, small-scale/artisanal, or recreational. Consumption of wild-caught fish is exceeding the capacity of this natural capital to regenerate. While estimates are controversial because the data used are often difficult to obtain or inaccurate, when last analyzed in 2005, almost a third of ocean fish species were collapsing (meaning that under 10 percent of the population remained), and it is estimated that *every* wild-caught species of fish will collapse by 2050.[13]

In aquaculture systems, aquatic organisms (e.g., algae, crustaceans, fish, mollusks, seaweed) are cultivated (thus the application of the term "farmed" to these organisms—e.g., "farmed fish"). Aquaculture systems may be coastal (near shore) or offshore. The primary ecosystem services derived from fisheries and aquaculture systems is the provisioning of food. Concerns about mercury and other contaminants (e.g., plastics, dioxins, and polychlorinated biphenyls [PCBs]) and animal rights aside, fish are considered to be a healthy dietary component because they supply protein and are a source of omega-3 essential fatty acids (deemed important for cardiovascular health). Fish constitute 7 percent of the world's protein consumption, 17 percent of the world's animal protein consumption,[14] and may constitute as much as 50 to 90 percent of individuals' animal protein consumption in island or coastal nations.[15] Human consumption represents the main use of fish (87 percent of fish utilization), with aquaculture supplying an increasingly large percentage (40 percent) of fish consumed over the last twenty years.

Aquaculture, including for seaweed, crustaceans, shellfish, and finned fish, is carried out within seascapes or using inland ponds and other artificial structures. Farming in seascapes generally uses cages and pens immersed in the water and can be done both by the shore and far offshore. Seascape fishing can also use a technique known as ranching, in which young fish are captured and confined in pens and allowed to grow until harvest. Cages and pens in aquaculture are generally porous to the environment and can lead to nutrient overloading of the surrounding habitat if fish are overstocked, and the high density of fish can lead to pathogen outbreaks that attack fish within and external to these enclosures.[16] Escapes of fish are also not uncommon, a recent example being the escape of 305,000 Atlantic salmon into the Pacific Ocean in Washington State.[17]

Aquaculture may seem like a way to provide an alternative seafood supply that would help with making wild-caught seafood sustainable by reducing demand, but sustainable aquaculture has proven difficult to achieve. Almost a fifth of wild-caught fish is used as feed for farmed fish; thus aquaculture can contribute to collapsing fisheries as much as open-ocean fisheries. There have also been associated negative environmental outcomes, such as pollution, freshwater contamination, habitat loss, the spread of exotic species, increases in the spread of seafood diseases, biodiversity loss, the production of low-quality food that sometimes contains toxins or microbial contaminants, and inequitable distributions of both income and food derived from aquaculture. Reminiscent of Green Revolution farming, the expanse of "Blue Revolution" aquaculture is currently unsustainable, but transitioning to sustainable aquaculture represents an important new horizon, much as sustainable terrestrial farming does. As in terrestrial systems, our management of marine systems is leading to dramatically less diverse systems as fish populations die out. With respect to regulating ecosystem services, fisheries may regulate climate via food web interactions that control primary production and therefore C cycling. For example, fish consuming zooplankton regulate zooplankton populations, thereby allowing phytoplankton (primary producers that use CO_2 for photosynthesis) to flourish. Aquaculture systems may also regulate water quality by enhancing nutrient removal (e.g., uptake of excess N in coastal waters), reducing turbidity via removal of suspended material by filter feeders, and trapping sediment. Coastal or nearshore aquaculture systems can regulate extreme events and protect coastal communities from storm surges. Aquaculture systems (e.g., seaweed cultivation, systems that use ponds) also demonstrate potential to serve as C sinks by storing C in sediments.

Culturally, fisheries and aquaculture can support social networks through community organizations, cooperatives, and the collective management of resources. These systems can also provide the educational benefits of aquatic resources management (e.g., management strategies, costs and benefits of aquaculture, consumption patterns and overfishing). Traditional (i.e., noncommercial, subsistence) fisheries provide such additional cultural benefits as spiritual and religious inspiration, support of traditional knowledge networks, and a sense of place.

SUSTAINABLY MANAGING ECOSYSTEM GOODS AND SERVICES

In the absence of management, ecosystems sustainably deliver multiple services. Management alters their composition, structure, and functions to increase their delivery of a desired good (or goods), but in doing so, all other goods and services change. In particular, regulating and supporting services decline, which translates into a loss of sustainability.

Fundamental Differences Between Natural and Agroecosystems

In comparing natural systems to their managed counterparts in agriculture, some fundamental observations emerge, which are illustrated in figure 2.3.

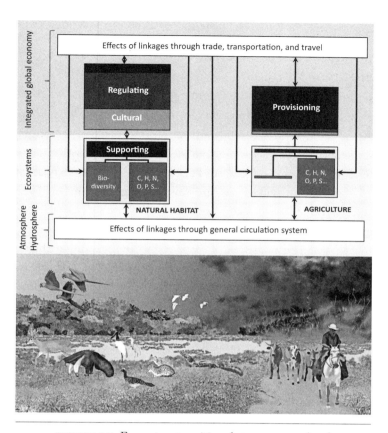

FIGURE 2.3 Ecosystem transitions between natural and managed systems.

Top: The left and right illustrate the four ecosystem service categories and biodiversity, adapted from the Millennium Assessment's framework (figure 2.1). Conversions of natural habitats (left) to agriculture (right) change the distribution of ecosystem services each provides. The usual transition is one in which provisioning services are increased but all other services decline. Bottom: To visualize what is presented in the top, this figure illustrates the conversion of a natural system to a grazing system. The Brazilian Pantanal, an extensive wetland, is rich in biodiversity. It provides regulating, cultural, and supporting services but little in the way of provisioning services (e.g., food). Large sections are being converted to cattle ranches, which provide considerable quantities of food, but at the expense of other services.

Source: Authors.

1. AGROECOSYSTEM COMPOSITION REFLECTS BIODIVERSITY LOSS

Only one or a few species dominate agroecosystems, whether terrestrial or aquatic. Interestingly, of the estimated 8.7 million species that have accompanied our existence since our origin some 200,000 years ago, only a small proportion (less than one-tenth of 1 percent) are farmed. For example, more than a quarter of all land devoted to agriculture is used primarily for raising livestock, yet that livestock consists of fewer than 40 species of vertebrates. Given that there are 40,000 species of vertebrates on Earth, that's a minuscule fraction of biodiversity. The remaining three-quarters of land used for farming includes over 7,000 species of cultivated plants, which seems like a lot, but only 30 of these provide 95 percent of our plant-derived food, and only four of these (rice, wheat, maize, and potato) supply more than 60 percent of this food. When one considers that there are 400,000 plant species on Earth, what we consider edible and have taken up in farming is a tiny proportion of Earth's biological diversity.[18]

Seafood farming is similar. Aquaculture, which has grown dramatically over the last 50 years, is estimated to supply half of the seafood we eat, but it focuses on a small number of species. Aquaculture cultivates approximately 300 species, but only about 20 account for roughly three-quarters of aquaculture production. What we don't cultivate we harvest from the ocean, which is home to some 20,000 species of fish. We don't know how many species we harvest, because 40 percent of what we catch consists of incidental or nontarget fish collectively referred to as *bycatch*. The FAO has records for about 1,000 species, so we eat only a small portion of the ocean's diversity. Of those 1,000 species, merely ten account for 27 percent of the fish we harvest, similar to how only a few plant species account for much of land-based agriculture.[19]

2. AGROECOSYSTEM STRUCTURE REFLECTS SIMPLIFICATION

In natural ecosystems, the consumer food web consists of herbivores (consumes plants), carnivores (consumes primarily animal matter), omnivores (consumes both plant and animal matter), frugivores (consumes fruit), granivores (consumes seeds), pollinators (consumes nectar and pollen), parasites, pathogens, and more. In agroecosystems, the consumer food web consists of plants and us as herbivores or us as omnivores and a few cultivated animal species. All unwanted herbivores and predators are considered pests. Unwanted plants are considered weeds. We strive to eliminate unwanted species and any species that is a pathogen or parasite of the species we cultivate. Compared to natural systems, agroecosystem food webs are simple. The decomposer food web, however, often retains considerable complexity. The decomposers are often hidden, generally found in the litter and duff on top of the soil, in the soil itself, or in the sediment of aquatic systems. These consist of hundreds of species of small animals, like worms, insects, mites, millipedes, and many other TAXA, as well as bacterial and fungal species numbering in the billions of individual cells per gram of soil. In farms, we manage decomposer communities indirectly by the way we manage soils: how we till, how we water (e.g., rain fed or irrigated), what nutrients we add, and what biocides we use. In some operations, decomposition is managed separately through composting operations in which farmers turn dead organic matter into compost that is rich in nutrients and returned to the soil. Though complex, agroecosystem decomposer food webs are still substantially lower in animal and microbial diversity than their ecosystem predecessors. *Simplification* in agriculture often refers to this simplification in composition and structure.

3. AGROECOSYSTEMS PROVIDE SUBSTANTIALLY INCREASED PROVISIONING SERVICES AND SUBSTANTIALLY DECREASED REGULATING AND SUPPORTING SERVICES

In the examples here and as illustrated in figure 2.3, natural systems provided all four classes of ecosystem services, but provisioning services could support only small populations of humans. Conversion to agroecosystems dramatically increased provisioning ecosystem services but was accompanied by dramatic reductions in regulating, supporting, and, often, cultural services.

4. NATURAL-TO-AGROECOSYSTEM TRANSFORMATIONS INITIALLY LEAD TO A DECLINE OR LOSS IN CULTURAL SERVICES, BUT THESE MAY BE REPLACED BY NEW CULTURAL SERVICES ASSOCIATED WITH MANAGED SYSTEMS

Because cultural services are tied to perceptions and beliefs, cultural values of managed ecosystems may replace those of unmanaged systems. The Brazilian Pantanal, for example, is a large, productive wetland in the center of South America, once populated by the Paiaguá tribe, consisting of hunter-gatherers who navigated the land using canoes and spiritually identified with a freshwater fish group known as pacu. European eradication of the Paiaguá diminished these cultural services because neither the Europeans nor others who replaced the Paiaguá valued the Pantanal in the same way. The Pantanal, however, is now home to 30 million cattle, and the Pantaneiro cowboys value the Pantanal as a grazing land that supports their horse- and livestock-based lifestyle. These grazing-land values, however, are now threatened by government proposals to drain the Pantanal for industrial soybean production. This example of the volatility of cultural services can be seen in most systems where indigenous peoples are threatened or declining or where the social and cultural mores of the residents must coevolve with changing political and economic trends.

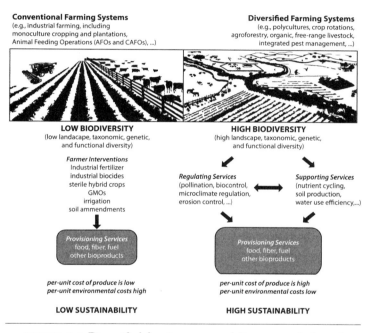

FIGURE 2.4 Diversified farming systems deliver a more balanced portfolio of ecosystem services.

Low-diversity farming systems, often typical of conventional, industrial farming operations, as illustrated on the left, require more farmer interventions and have higher environmental impacts, but production is often more cost effective. Diversified farming systems, such as shown on the right, require fewer interventions, provide a wider array of ecosystem services, and have lower environmental impacts, but the cost of production is often high.

Source: Authors.

5. AGROECOSYSTEMS, DEPENDING ON MANAGEMENT, CAN PROVIDE SIGNIFICANT REGULATING, SUPPORTING, AND CULTURAL SERVICES

Currently, agroecosystem management diminishes climate, hydrologic, pollination, and other regulating ecosystem services, but these losses arise primarily because of the practices managers employ. Cropping schemes, tillage, biocide use, and irrigation can either protect or erode ecosystem service delivery. Monocultures and CAFOs are examples of agricultural management in which virtually all ecosystem services are lost; by contrast, integrated and diversified farming systems (figure 2.4) can optimize multiple service delivery.

ECOSYSTEM SERVICE VALUATION

One of the motivating factors behind increased interests in ecosystem services concerns the possibility that their inclusion in our economies could further sustainable development. We market provisioning services and, to a certain extent, cultural services, but neither regulating nor supporting services yet have markets. No one pays for primary production, decomposition, nutrient cycling, or the generation and maintenance of biodiversity by evolutionary processes. If all ecosystem goods and services were included in our economies, however, transitioning from traditional, unsustainable development to the new paradigm of sustainable development could accelerate.

A controversial study by Costanza et al. published in 1997 and updated in 2014[20] attempted to assess the value of ecosystem services using conventional methods in economics to answer the question: What is the value of all the services that ecosystems provide for us in a single year? The updated estimate in 2014 (using 2011 data) was that it would be worth $124.8 trillion dollars. In other words, if ecosystems ceased delivering services for one year, the cost to us would be $124.8 trillion. Right now, they are delivered to us free of charge, which is like receiving an annual gift of $124.8 trillion from the world's ecosystems. Considering that the annual global gross domestic product (GDP), or the annual global marketed value (in monetary terms) of *all* goods and services, is around $75 trillion, it means that what ecosystems provide us in terms of unmarketed goods and services is substantially greater (almost double) than our world economy. We should reiterate that these studies are controversial, but they do provoke us to think about the issues.

Less controversial is the work of TEEB, whose goal is explicitly to make nature's values more visible to the public and to policy and decision makers. Although global in scale, the specific analyses are primarily done at national or subnational scales, which provide for results that may be more relatable to individuals. Some examples of published analyses they have highlighted are:

- Conserving forests avoids greenhouse gas emissions worth US$ 3.7 trillion
- Global fisheries underperform by US$ 50 billion annually
- Beekeeping generates US$ 213 million annually in Switzerland
- Tree planting benefits to urban dwellers in Canberra, Australia, are valued at US$ 20–67 million over the period 2008–2012
- Pollination services provided by forests in Sulawesi, Indonesia, are valued at US$ 53 per hectare ($22 per acre)
- Water provisioning from the tropical-forest Leuser Ecosystem in Indonesia is valued at US$ 2.42 billion

- Climate regulation by tropical forests in Cameroon is valued at US$ 842–2,265 per hectare ($341–917 per acre) per year

Delineated by specific service, these findings are more digestible than Costanza et al.'s conclusion that all the world's ecosystem services are worth $124.8 trillion.

TEEB's approach of vetting, aggregating, and promoting such valuations may be more palatable to many than Costanza et al.'s approach, but either way, they both make the salient point that the inclusion of ecosystem services in how we value managed and unmanaged ecosystem goods and services is important for advancing our transition to environmental sustainability.

More importantly, TEEB for Agriculture and Food has issued a series of reports that provides methods, analyses, and promotes true or full accounting of the costs and benefits of food systems.[21] It models food and agriculture as three interconnected systems: human systems, agriculture and food systems, and natural systems. Take, for example, the argument that the dominant portion of the human population, the urban and rural poor, needs cheap food that is primarily produced by conventional or industrial methods. TEEB finds that this is not "cheap" when the health and environmental costs are included. One of its studies explored the production costs and environmental costs of maize agriculture in Mexico, Ecuador, and the United States. The report found that in Mexico the "shadow price" of rain-fed maize grown for self-consumption in 2011 was nineteen times higher than the market price for commercially produced maize grain for human consumption (largely white maize, as opposed to yellow maize, which is used for animal feed). The striking difference was attributable to the greater adaptability to variation in environments in the region; lower losses to pathogens and pests; and the cultural, spiritual, religious, and culinary values associated with the huge diversity of the landraces of the rain-fed, self-consumed maize.

SUMMARY

Farming modifies the characteristics of natural ecosystems, such as their species composition, structure, and biogeochemistry, to maximize the production of goods we value and for which there are markets. When we manage a forest as a plantation to produce citrus, a rangeland as a grazing system to produce beef, or a coastal habitat as aquaculture to produce shrimp, we have converted natural systems to agroecosystems. These gains in production of valued goods, however, are almost always accompanied by severe losses in ecosystem services, such as soil formation, climate regulation, freshwater production, and pollination. Ecosystem services have tremendous value, but because they are unmarketed, their losses during agricultural development goes unchecked—and the resulting systems become unsustainable. Such losses in ecosystem services, however, are often neither intentional nor necessary, and with the right agricultural practices they can be minimized or avoided altogether. To do so, however, requires significant social change. Markets for ecosystem services need to be generated, managing portfolios of ecosystem services needs to be incentivized, and ecosystem services need to be included in agricultural and environmental policies. Modern environmental and ecological economics attempt to facilitate such change via quantitative valuations of ecosystem services, often in monetary terms. Thus far, these analyses demonstrate that unmarketed ecosystem services often far exceed the value of marketed agricultural goods. Such findings help make the importance of ecosystem services more visible to the public and to policy makers and decision makers. As the inclusion of ecosystem services becomes more prominent in guiding management and shaping policy, food, farming, and agriculture will become more sustainable.

3

SUSTAINABLE DEVELOPMENT AND FOOD PRODUCTION

LINKING THE THREE PILLARS OF SUSTAINABLE DEVELOPMENT WITH FOOD PRODUCTION AND FARMING

This chapter builds upon the principles of environmental sustainability we reviewed in chapter 1 and the ecosystem service framework we reviewed in chapter 2 to explore each of the three dimensions, or PILLARS OF SUSTAINABLE DEVELOPMENT, as they relate to food production. Here, we consider the environmental, social, and economic dimensions of sustainable development separately. Of course, as chapters 1 and 2 emphasized, sustainable development integrates across these three dimensions, and, as we will see, they are intrinsically related.

ENVIRONMENTAL SUSTAINABILITY AND FOOD PRODUCTION

The steady state of nature is so familiar to us all that we seldom think about it. We expect that repeated visits to a forest, prairie, or coral reef will reveal a system that varies little from year to year or even from decade to decade. And we are generally right; in the timescale of a typical human lifetime, nature changes little. Even in our current epoch of the ANTHROPOCENE, characterized by rates of change unprecedented in Earth's history (except for times of mass extinction), significant environmental change still takes decades. In agricultural systems, however, if management were to cease, change would be rapid—domestic species would crash, and exotic or native species would take over. The central theme in this section is that nature is highly dynamic over small spatial and temporal scales but intrinsically sustainable over larger scales. The sustainability of managed systems, at all scales, is in our hands.

Ecological Balance

A simple truism: Every species, be they plant, animal, or microorganism, must lead life sustainably or perish. Consume more than a system produces, and eventually you or your system—or both—will collapse. In natural ecological systems, countless processes keep production and consumption balanced. If a soil microbial community grows and consumes the majority of the available nitrogen in the soil, the reduced fertility of the soil will reduce vegetation. Reduction in vegetation leads to a reduction in carbon input into the soil, which in turn will starve the microbes that depend on vegetation for their carbon, and the end result is that the system swings back to a more balanced state. Likewise, if a grazer overgrazes, if a predator overdepredates, if a parasite debilitates its host, if a flower does not reward its pollinator, or if any species does excessive harm to those it depends on, it will either decline or disappear. Over the short term, populations fluctuate in nature, but over the long term, these sorts of feedbacks in ecosystems link species to one another and ensure ECOLOGICAL BALANCE.

Agroecosystems are quite different—we control the feedbacks with management. If pests threaten an agroecosystem's function, we use biocides. If precipitation is low, we irrigate. If soil nutrients are depleted, we add fertilizers. If parasites or microbes threaten livestock, we medicate. If decomposition is too slow, we employ composting techniques to hasten the process. If fishery populations are shrinking, we impose catch limits.

Ecological Stability

Sustainability is not just about balancing consumption and production; it's also about ECOLOGICAL STABILITY. Environments vary constantly, but rarely in ways that imperil species or system function. Sometimes, however, environmental variation exceeds its norms, creating droughts, floods, fires, hurricanes, or other extreme events that significantly change an ecosystem's composition or alter its functions and services. When such extreme events occur, also known as environmental perturbations, disturbances, or shocks, stability becomes important. Systems either survive shocks or they collapse.

Natural systems are well adapted to environmental variation, but again, whether agroecosystems can survive environmental shocks is a matter of management. What crops we plant, how heavily we fish, how densely we stock a rangeland, or how much we pressure an agroecosystem to produce will determine its stability. If external factors are extreme, however, like a severe drought, a rise in temperature that plants cannot endure, or the emergence of a virulent and untreatable livestock pathogen, there may be little that management can do to prevent agroecosystem collapse.

Balancing Inputs and Outputs

In chapters 1 and 2, we considered that no ecosystem exists in isolation. Ecosystems, managed or unmanaged, are essentially materially open systems that are contained by a materially closed biosphere. By open, we mean that ecosystems constantly capture and lose inorganic and organic material. Plants can capture carbon from the atmosphere, nitrogen-fixing bacteria can similarly capture atmospheric nitrogen, and salmon can migrate from oceans upstream to spawning grounds where they perish and decompose, importing ocean-derived nutrients into terrestrial systems. Conversely, soil can lose nutrients to groundwater and runoff that makes its way to the sea; plants, animals, and microorganisms respire water and carbon dioxide (CO_2), which enters the atmosphere.

Like the simple truism that production and consumption must balance over the long term, system inputs and outputs must also balance over the long term. If outputs consistently exceed inputs, the system shrinks away. If inputs exceed outputs, the system grows until external sources of inputs are depleted.

Agroecosystems, like any ecosystem, have inputs and outputs, both of which determine production and sustainability. Anthropic, or human-based, inputs and outputs are distinct from those of natural systems. When agriculture started, agroecosystems were small in scale and likely differed little from other systems, experiencing inputs and outputs much like any other. Today, however, there is a gradient from rural subsistence farms, which have little in the way of managed inputs and outputs, to industrial farms that may have significant inputs imported from well outside their boundaries and outputs exported to virtually anywhere in the world.

Productivity and Hidden Inputs

Typically, agricultural productivity, from an economic perspective (i.e., not yields), is calculated using the ratio of the market value of the output to the market value of the input. For example, if two farms input the same quantity of synthetic fertilizers in terms of pounds of nitrogen per acre and have the same yields in terms of nitrogen content of harvested bioproducts, the farm that paid less for its fertilizer would be considered to have greater productivity, even though, in terms of nitrogen output and input, the two farms had exactly the same ratio. In this way we can see that biophysical measures are not usually included in economic calculations of productivity.

Nonmarketed, or *hidden*, inputs and outputs are often not included in the ratio, so estimates of production are often distorted. Hidden inputs and outputs range from tangible, familiar things like greenhouse gases and pollutants to less tangible things, such as social factors like traditional knowledge or economic factors like government subsidies. The concepts of natural capital (chapter 1) and ecosystem services (chapter 2) promote a fuller accounting of inputs and outputs, but so long as many ecosystem services and hidden inputs and outputs are not marketed, agronomic accounting cannot provide an accurate picture of the ratio of outputs to inputs. For example, pollination by bats worldwide was estimated to be worth over $200 billion in 2005, which was 9.5 percent of the value of world food crops that year. In another example, bat control of crop pests in Nuevo León, Mexico, was valued at $479,000 to $1.2 million per year.[1] For farms where bat pollination was important, failure to include these intangible regulatory ecosystem services in estimating output-input ratios means that farmers had a false, inflated sense of productivity. That is, their output-input ratio was higher than it would be if we corrected for hidden ecosystem service inputs that would cost farmers were they to be eliminated.

One can see that sustainable agricultural practices are those that keep the output-input ratios high. Organic farming, no-till farming, the use of cover crops, integrated pest management, crop rotation, livestock-crop integration, livestock diversification, agroforestry, and water conservation are all examples of practices that reflect the strategy of optimizing or maximizing the output-input ratio.

Stability

A PLETHORA OF TERMS

We generally think of environmental sustainability as stable functioning without harmful environmental impacts. The term *stable*, however, is tricky. STABILITY is a catch-all term for a variety of different ways of thinking about a system's dynamics, which we summarize in box 3.1.

Ecosystem stability, in its most basic sense, is simply the ability of an ecosystem to remain relatively unchanged over time. Our environment is highly variable, showing patterns in spatial and temporal variability that range from small and manageable (e.g., daily temperature fluctuations) to larger (e.g., interannual shifts in climate across the globe associated with El Niño and La Niña events) to huge and catastrophic (e.g., 10,000-year glaciation cycles).

Ecosystems can respond to environmental variation, including significant deviations from norms, in a variety of ways. Natural systems are often relatively stable, rarely exhibiting major changes over the long term unless exposed to environmental shocks. Agroecosystem stability, however, is a function of management. In general, farms are inherently unstable in the sense that if human interventions cease, they return to their natural states. A farm's return to its natural or prefarming state, what ecologists call SUCCESSION, can be a slow process. Evidence of Mayan agriculture, which took place two to three thousand years ago, can still be found in the sediments and soils of Mexico's Yucatan Peninsula.

What's interesting and important from the standpoint of food production is understanding what makes a system stable and whether its stability is fragile and easily perturbed or robust and capable of enduring environmental change. Imagine a system in which sunlight, precipitation, and temperature are constant; inputs and outputs are always balanced; and every population of plant, animal, and microorganism never increases or decreases in size. Such an imaginary system would meet our definition of stable. What if, however, there were a DISTURBANCE, shock, or PERTURBATION to the system, such as a drought, fire, flood, invasion by an exotic pest, or loss of a species from a lethal disease? In the face of disturbance, there are several questions we would want to know the answers to. Can the system remain constant or *resist* change in the face of perturbation? If the system cannot resist change, is it resilient? Do the system's abilities to resist or be resilient to change depend on the magnitude of the disturbance? Was the change minor (some populations and production declined a bit), or was there massive change (a collapse)? Did the system show no change for a long time even though there was chronic environmental change, like steadily increasing temperature, but then all of a sudden collapse, as if it crossed some kind of THRESHOLD/TIPPING POINT? Or did the system undergo radical change, never to return to the original state again, but instead assuming some new state with different ecosystem properties, creating what is referred to as a REGIME SHIFT to an ALTERNATE STABLE STATE?

BOX 3.1 ECOSYSTEM STABILITY

Stability, though often focused on resilience, is a central objective in sustainable development and sustainable food production. Stability, however, is a complex, abstract, often mathematical concept that generally refers to the dynamics of any system. Here, we define the terms associated with stability in the context of ecosystems.

State variables: ecosystem properties (box 2.1), such as number of species, relative abundance of species, nutrient content, area occupied, and carbon stored

SYSTEM STATE: specific set of values at a given point in time for a system's state variables

SYSTEM DYNAMICS: properties of temporal changes in state variables

Linear dynamics: dynamics are predictable or proportional to changes in environmental conditions

Nonlinear dynamics: dynamics are difficult to predict, seemingly chaotic, and not always proportional to changes in environmental conditions

Chaotic dynamics: nonlinear dynamics, more specifically referring to aperiodic, seemingly random fluctuations in state variables in which there is little certainty in near- or long-term states

Variability: the range of values, usually bounded by upper and lower limits, that the state variables can have over time

Perturbation, disturbance, shock: significant deviation from norms in environmental conditions, e.g., a drought, flood, or major fire

Stability: change in state variables over time (the lower the change, the greater the stability)

STABLE STATES: sets of different states a system can exhibit that are stable

Resilience: time it takes to return to a preperturbation state

RESISTANCE: extent to which a system deviates from its original state when perturbed

Reliability: the probability of occupying a specific state over a unit of time (e.g., a maize yield reliability of 0.5 for 8 tons of maize per hectare [127 bushels per acre] per year would mean that the probability of obtaining 8 tons per hectare in any year is 50 percent)

PERSISTENCE: length of time a system remains within state boundaries independent of other stability properties

CATASTROPHIC CHANGE, regime shifts, and COLLAPSE: system shift to an alternative stable state, often rapidly and with little warning

Thresholds and tipping points: the specific set of conditions describing a threshold that, if crossed, would trigger a catastrophic change, regime shift, or system collapse. Most ecosystems have thresholds and tipping points, but they are generally unknowable and can only be approximated.

THRESHOLDS AND TIPPING POINTS

Of the many ways ecosystems differ in their stability, none are as worrisome as their thresholds and tipping points. An ecological threshold is a particular environmental condition, like a specific temperature or level of grazing, at which a system undergoes a rapid change to an alternative stable state. Because the system is frequently envisioned as teetering on an edge just before it tips over and falls or switches over into an alternative stable state, the point at which the threshold lies is called the *tipping point* (figure 3.1).

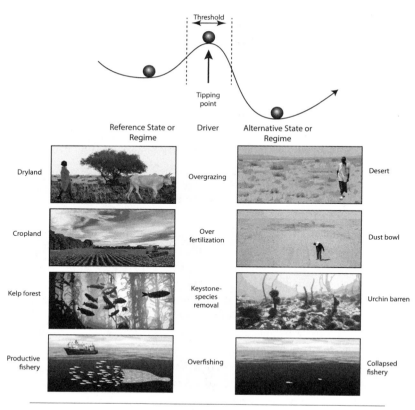

FIGURE 3.1 Thresholds and tipping points.

Thresholds are often illustrated as the top of a hill separating two valleys, each representing a different state. A ball, representing the system, rests at the bottom of a valley but changes its position when forces (drivers) move it up to the top of the hill. If pushed over the point where it rolls down into the adjacent valley (i.e., pushed over the tipping point), it comes to rest in a different, alternative state. The threshold itself is a range over the hill's top, within which a tipping point exists. The tipping point, however, is seldom precisely locatable. To illustrate the concept, four ecosystems, a dryland, cropland, kelp forest, and fairly productive open-ocean capture fishery, are shown on the left. Each can shift to an alternative state when experiencing different drivers, such as overgrazing, overfertilization, the removal of keystone species, or overfishing. The alternative states, a desert, dustbowl, urchin barren, and collapsed fishery, are shown on the right. As the diagram at the top suggests, while one can drive a system from its alternative state back to its original state, the necessary effort (pushing the ball up the long, steep slope of the valley on the right) may be much greater than what it took (pushing the ball up a shorter, less steep slope of the valley on the left) to cause the state transition in the first place.

Source: Authors.

We often know *what* causes tipping points but do not know precisely *where* the point is. We know, for example, that if we chronically overfish a species, its population will collapse, but at what specific level of fishing the fishery will collapse is difficult to know and has a certain amount of randomness to it. The collapse of the Northwest Atlantic cod fishery in 1992 is an example. Fishing pressure continuously increased with no particular signs that a collapse was imminent. Then, in 1992, the catch drastically declined to 1 percent of the previous year.[2] The Canadian government called for a moratorium on Atlantic cod fishing, but it took a decade before populations began to show signs of recovery. To this day, they remain at dangerously low levels.

Drylands similarly show threshold responses to grazing pressure. Grazing can continuously increase without warning that a collapse is imminent. Drylands are stable, sustainable states, but they have thresholds for grazing pressure that, if crossed, lead to a high probability of desertification. Poverty, hunger, and conflict usually follow such an event.[3]

Wherever thresholds and tipping points have been observed, they were associated with a chronic or persistent environmental change, often referred to as STRESSOR, PRESSURE, or DRIVER (e.g., fishing, grazing, pollution, or rising temperature). The driver causes one or more ecosystem state variables to respond slowly, at first, but eventually the rate of change accelerates dramatically.

Tipping points are worrisome at the ecosystem level but even more so at the planetary level. Currently, we are experiencing several planetary-scale, chronic environmental changes, including climate change, mass extinction, nitrogen pollution, increasing ecological invasions across all ecosystems, and widespread emerging and reemerging disease. Neither unmanaged nor managed systems are coping well with these co-occurring planetary-scale changes. The idea of tipping points motivated the development of the somewhat controversial planetary boundaries framework (chapter 1),[4] which addresses the concern that Earth's state variables, whose current values are conducive to humanity, may with little warning change dramatically. Global food production, for example, now at levels that can support humanity, are also sources of the same chronic environmental stressors just listed. If there are planetary-scale tipping points, there could be a global regime shift in which food supplies crash worldwide.

Land Use and Land Management

How we use and manage land for food production governs the emission of GHGs, soil erosion, water use efficiency, biodiversity loss, and other factors that can contribute to land degradation. Land use and management often focus on farms, but farms reside in LANDSCAPES, which are generally defined as a group of interacting ecosystems within a specified region. Thus, how an agroecosystem is managed is important for sustaining production on a given farm as well as for how it affects the surrounding environment through its interactions with neighboring systems.

At the landscape level, perhaps the most prominent management issue concerns LAND SHARING versus LAND SPARING,[5] a framework that explores tradeoffs between increasing agricultural yields in agroecosystems by intensification versus increasing agricultural yields by expanding land use and land transformations but using less intensive agricultural methods. Both have adverse environmental consequences—intensification, or industrial agriculture, has numerous adverse environmental impacts and poses major challenges to achieving sustainability because such agroecosystem transformation leads to large losses in nonmarketed ecosystem services, as described in chapter 2. If one intensifies agricultural production, yields increase and land is spared, but if one expands agroecosystems to increase production at the landscape level, then the share of land that goes to farming increases. The adverse impacts of land sharing can be minimized by employing sustainable land-management practices, such as diversified farming (chapter 2). The adverse impacts of land sparing through intensification can also be minimized if intensification does not result in additional harmful environmental effects.

The land sparing/land sharing framework is seen by some as too simplistic in dichotomizing land types into agricultural and natural, focusing on yields, and ignoring efforts to restore degraded ecosystems across the landscape. Governance, land scarcity, food distribution, and numerous social factors, such as livelihood strategies, poverty, and equality, need to figure into the framework.[6] For example, one of the major concerns over the expansion of agricultural lands is deforestation, and rather than using the sparing/sharing debate focused on production, policy makers, scientists, practitioners, and producers have collectively developed land-use policies and programs to reduce deforestation and environmental degradation while supporting both human and economic development. For example, the UNITED NATIONS REDUCING EMISSIONS FROM DEFORESTATION AND FOREST DEGRADATION IN DEVELOPING COUNTRIES (REDD+) program represents one such effort to reduce deforestation while supporting sustainable development. Similarly, the United Nations Convention to Combat Desertification focuses on the complex social and economic factors that drive desertification in dryland rather than focusing on yields.

Climate-Smart Agriculture

When it comes to climate change, one strategy, if we can call it that, is to do nothing. When social and natural scientists of the UN Intergovernmental Panel on Climate Change develop scenarios for the future based on such a *business-as-usual* approach, they predict worldwide insufficiency and instability in food supplies that will harm billions, especially the poor and vulnerable. We already face a major challenge in finding ways to sustainably increase agricultural production to feed 10 billion people by 2050, but this will prove impossible if we don't address climate change.

The opposite of doing nothing when confronted with a serious challenge is to be *smart*, which means doing something to address the challenge. CLIMATE-SMART AGRICULTURE is agriculture that does just that—it actively addresses climate (box 3.2). Climate-smart agriculture was only formally described by the Food and Agriculture Organization of the United Nations (FAO) in 2010;[7] it has since gained considerable traction as a sustainable development strategy. It aims specifically to sustainably increase agricultural productivity, enhance both agricultural and social resilience to climate change, and mitigate climate change (i.e., reduce greenhouse gas emissions).[8]

There is an ever-growing range of climate-smart practices, most of which are common to sustainable agriculture (box 3.2). Such climate-smart agricultural practices are fundamental to sustainable agriculture, but they are even more beneficial in the face of climate change.

One last important feature of climate-smart agriculture is its emphasis on flexibility. We know the climate is changing, but precisely when and where is difficult to know given the huge number of variables. Ecosystem stability and tipping points,

BOX 3.2 CLIMATE-SMART AGRICULTURE

Climate-smart practices, most of which are common to sustainable agriculture, include:

1. Improving agroecosystem resilience in the face of increasing climate variability by
   a. increasing organic matter inputs to soils to improve a soil's water holding capacity, reduce runoff and erosion, and increase nutrient availability,
   b. terracing to reduce runoff and soil erosion and improve water drainage,
   c. using drought-tolerant plant varieties where the incidence and severity of dry spells is rising,
   d. using water harvesting/rainfall capture methods for irrigation during dry spells and to reduce water runoff and soil erosion, and
   e. reducing tillage, also to prevent runoff and erosion.
2. Reducing greenhouse gas emissions by
   a. improving inorganic nitrogen fertilizer management,
   b. improving livestock waste management, and
   c. incorporating renewable energy sources (e.g., solar-powered irrigation pumps) to reduce reliance on fossil fuels.
3. Improve carbon sequestration and storage by
   a. use of agroforestry systems,
   b. reduced grazing intensity and grass cultivation in pastures in arid or dryland systems or by frequent movement in systems where precipitation is adequate,
   c. incorporating perennial crops to enhance carbon sequestration, and
   d. reducing tillage to maintain soil structure, enhance the accumulation of organic matter, and reduce water runoff and soil erosion.

spatial and temporal variability, the ever-changing face of climate policy, enormous variability in people's responses to climate change (from climate-change deniers to highly motivated environmental activists), and the constant emergence of new information make climate-smart agriculture a practice that cannot be rigid or a one-size-fits-all prescription. Examples of climate-smart agriculture flexibility include

- providing guidelines for crop choices based on climate models, not on regional traditions;
- land-use policy incorporating climate change, such as lowering stocking rates according to changes in precipitation predicted by climate models to prevent drylands from reaching desertification tipping points;
- providing climate-change insurance or climate-smart subsidies for farmers; and
- continuously updating climate-change information resources for farmers, fishers, ranchers, investors, insurance providers, extension agents, policy makers, and others in the agricultural sector so they can make climate-smart decisions.

As climate-smart agriculture becomes increasingly prevalent, farmers will be able both to mitigate and adapt to climate change.

SOCIAL SUSTAINABILITY AND FOOD PRODUCTION

Sustainable production is as much about social sustainability as it is about environmental sustainability. In this section, we explore these social factors as they relate to food production. We group them into four traditional categories: social factors that concern (1) equity, (2) awareness, (3) participation, and (4) SOCIAL COHESION.

Equity

Equity concerns fairness, impartiality, justice, and other complex aspects of how people treat one another. When it comes to food production, the equitability of access (to food by consumers and to land by farmers) constitutes two key issues in sustainability.

Because most people do not grow their own food, consumption is determined by income and equity in access to food. For example, over half (55 percent) of the global population live in cities and do not grow their food. The distribution, quantity, quality, and cost of food in urban environments, therefore, lead to uneven distribution and inequity in food access. The urban poor often have lower access to healthy foods in their neighborhoods compared to those in higher-income neighborhoods. Globally, there are over 800 million urban poor; such inequity generates enormous skews in food consumption that biases agriculture toward the production of less expensive, nutrient-poor crops grown for profit.

The equity of food access affects the urban poor who do not grow their own food; in contrast, land inequity affects the majority of those who do grow their own food. Most farmers worldwide are poor, smallholder farmers or farm workers whose livelihoods are dependent on land access. While 30 percent of people are engaged in agricultural work worldwide (a number that varies widely by region: for instance, in sub-Saharan Africa, this number increases to 60 percent, while in the developed world it can be as low as 1 percent), the total amount of food production they control is minuscule.[9] The majority of the world's farms (72 percent) are very small, less than 2.5 acres (1 ha) in size, and another 12 percent are small, between 2.5 and 5 acres (1–2 ha). Yet, though small and very small farms collectively represent 84 percent of the world's 570 million farms, they cover only 12 percent of the world's farmland. In sharp contrast, large farms (larger than 124 acres, or 50 ha) represent an extreme minority (1 percent) of farms in the world, yet they control 65 percent of the world's farmland. As one can imagine, if your farm holding is small, the quality and yield of production is strongly dependent on the quality of your land. High-quality land, however, is expensive, often favoring wealthier farmers or agribusiness, who largely employ industrial farming practices. Furthermore, large farms have greater resources to endure climate change or extreme weather events than small farms.[10] Thus, one can argue that the food security of the urban poor is dependent on large-scale industrial farming.

These inequities in food access common to the urban poor and inequities in land distribution common to the rural poor collectively bias farming toward cheap, low-quality products that are invariably produced by typically unsustainable practices such as industrial farming. Because of this and in tandem with the rise in industrial food production over the past several decades, as discussed in chapter 1, social movements to promote equity in food and agriculture have emerged around the world.

Awareness

Most people, especially those who live in cities or far from farms, are often unaware of how food is produced and how it affects our environment, our economies, and people's livelihoods. Even people who work in agriculture may not be entirely aware of how our food system functions, as they may work in one part of the system that connects farmers to consumers (e.g., they may work solely in slaughterhouses, processing, shipping, or marketing). Additionally, a person's awareness differs across geographic location and socioeconomic status. If awareness of food-production processes and policies is poor among both consumers and farmers, these groups are unlikely to make choices that would improve food sustainability.

A number of organizations have arisen to promote food awareness and improve food literacy.[11] Greenpeace, Bioversity International, the Center for Food Safety, the GoodPlanet Foundation, and Food and Water Watch are just a few of the many organizations that strive to improve public awareness of food sustainability.[12]

Participation

Farming is rarely fully isolated, subsistence agriculture. Instead, it is almost always part of a group effort. A farm may be smaller, consisting of a single family or household, or it may be a larger enterprise that includes the landowner and tenants, a farmer and their farmworkers and employees, or an agribusiness that includes farmers, laborers, packagers, processors, transporters, and more. In every case, the sustainability of its production will be strongly influenced by the nature of participation among its members. Here we will focus on two key aspects of participation: the degree to which ownership is shared by workers and the community and the degree to which agribusinesses integrate their operations.

LAND TENURE AND PARTICIPATION

Individuals are more likely to participate and care for the functioning and success of an operation when they share ownership in it. However, the degree to which farmers and laborers share farm or land ownership varies enormously. Those who labor on a farm often do not own the land they work. For example, the *adhiya* in traditional Hindu societies is a system in which people of a lower social standing have access to land owned by people in a higher social standing but must share 50 percent of their profits.[13] Other examples include *sharecropping*, in which tenants give portions of their produce to the landowners, and *leasing*, in which someone temporarily uses the land for a fee. Such arrangements are useful for individuals who cannot afford the initial capital investment to purchase land, but they do not incentivize the kind of long-term management practices that favor sustainability.

In our discussions on equity, we noted that the vast majority of farms are small and farmed by the rural poor. These farms often lack the capacity to focus on sustainability. On the other hand, the majority of land around the world consists of large, industrial farms owned by a small fraction of the world's farmers and whose focus is often on production and revenue generation rather than sustainability. An alternative land tenure system would be one where farmland is shared by all those who work the land. The *ejido* in Mexico, as an example, is one where the agrarian population formed collectives to use common land for farming. This was once common in Mexico (in particular Chiapas) and led to a cooperative culture concerned with land health and sustainability as well as production. However, this form of farming was forcefully phased out as the Mexican government favored more production-oriented industrial farming in the 1970s. The *ejido* is a classic example of the tensions between the dominant, large-scale, production-oriented privately held farming systems and shared-ownership models. Community land trusts, as another example, are lands owned by nonprofits that are entrusted to communities to improve their collective well-being.[14] There are also partnerships, corporations, and limited liability companies that form legally binding cooperative agreements among farmers who share their land in a cooperative enterprise. Such alternative land tenure systems strive to encourage land stewardship and promote sustainable production.

BUSINESSES AND PARTICIPATION

Participation in businesses ranges widely from a single farmer selling directly to a small number of consumers to transnational agribusinesses employing tens of thousands of workers and selling to hundreds of thousands of consumers through global markets. Commitments to sustainability, not surprisingly, also range widely depending on the size of the business, access to capital, and ideology.

As we observed in chapter 2, there are many links in the SUPPLY CHAIN, including storage, processing, packaging, transporting, distributing, obtaining food safety clearances and certifications, financial and legal firms to deal with import/export rules and regulations, customs, tariffs, taxes, credit, loans, payments, marketing agencies, and, finally, the retailers where the customers purchase the farm goods. No matter how large the operation, unless the farmer sells directly to the end consumer, a farm either has to allocate some portion of its economic gains to pay the service providers at each link in the supply chain whose services they retain or sell directly to a buyer who then adds their markup and sees the product through to the next link in the supply chain or through all the links to the consumer. In general, the economics of supply chains means farmers rarely obtain the market value of their products.

In some cases, rural farmers need only access to a nearby local market to sell their goods to consumers. COMMUNITY-SUPPORTED AGRICULTURE is another means of more directly connecting farmers with consumers by creating contractual agreements in which the farmer provides shares of production to shareholders or subscribers, who are usually members of a local community.[15] Thus, community-supported agriculture represents an effort to shorten the supply chain.

Increasingly, some farms market their products via internet commerce, either directly, if they have the capacity to do so, or through sales or contracts with internet food businesses. Even small-scale farming operations can now have access to international markets, though these are invariably niche markets in which the products are unique and highly valued.

Large agribusinesses often invest their assets in acquiring or controlling the supply chain, a practice known as VERTICAL INTEGRATION. In so doing, they often sign contracts for the output and thereby dissociate from the most risky part of the supply chain: the production. BACKWARD VERTICAL INTEGRATION involves gaining control of links in the supply chain connecting the farm to the agribusiness. FORWARD VERTICAL INTEGRATION is the converse; gaining control of the supply chain links that lead to the consumer. Starbucks, for example, is a transnational coffee company that vertically integrates both backward, by contracting with coffee growers and owning its own roasting plants, warehouses, and distribution facilities, and forward, by selling directly to consumers through a network of 24,000 retail outlets in 70 countries.[16]

Agribusinesses, however, no matter their scale, can pursue sustainability if their participants decide to do so. A corporate board whose membership subscribes to the principles of sustainable development, an employee base that favors sustainable practice, or a consumer base whose preferences are for sustainably produced foods can all, in principle, lead or pressure an agribusiness to pursue sustainability. The World Business Council for Sustainable Development (WBCSD), for example, includes corporate participants like Archer Daniels Midland, Cargill, Kellogg, Monsanto, Nestlé, and PepsiCo, yet these transnational agribusinesses are considered some of the world leaders in industrial agriculture and sometimes considered synonymous with unsustainable agriculture.[17] Their participation in the WBCSD, whether genuine or as a matter of good public relations, may at least gesture to an awareness of sustainability, even if they show little evidence of corporate will to implement sustainable solutions. These large agribusinesses are not representative of many WBCSD members who do, in fact, pursue sustainability where possible.

Social Cohesion

Social cohesion is a complex concept that weaves together ideas about equality and social capital. The greater the equality of opportunities and access to the goods and services necessary for leading a good life, such as food, education, and health services, the more inclusive and cohesive a society is. Likewise, the stronger and more positive the relations, interactions, and ties among people are, the greater the social capital and cohesion among people. Social cohesion is an important component for sustainable farming. In particular, it is important for the intergenerational and community transfer of farming knowledge, in which knowledge of a particular agroecosystem (e.g., soil, climate, pest, best management practices, and best elements of social structure and interactions) is passed down from generation to generation and neighbor to neighbor, providing for the kind of long-term and landscape-driven perspective necessary for sustainability. This knowledge, however, can disappear within one generation if the intergenerational link is disrupted, and landscape-level management can suffer without farmer-to-farmer communication.[18] Since the mid-twentieth century, reliance on intergenerational local knowledge transfer has become less prevalent. In its place, an often bewildering array of science-based information sources on farming, food, diet, health, climate, and environmental issues (e.g., the pros and cons of GMOs, synthetic fertilizers, or biocide use) are made available or distributed to farmers and consumers. These sources include agricultural extension workers, schools and institutes of higher learning, nongovernmental (NGO) or civil society (CSO) organizations, governmental agencies, the World Bank, the FAO, private companies, and more. These forms of science-based knowledge dissemination and acquisition lack the social cohesion that traditional knowledge provided. Ensuring or restoring lost intergenerational social cohesion in farming communities and ensuring that contemporary information sources on food and farming emphasize sustainability are important elements of and for social sustainability.

Recently, more NGOs and grassroots movements, such as those described in chapter 2, including Campesino-a-Campesino, la Via Campesina, Movimiento dos Trabalhadores Rurais Sem Terra, the National Young Farmers Coalition, and the Association for India's Development,[19] support smallholder or independent farmers through knowledge sharing and social support. They often aim to modify national policies and advocate for farmer rights on the international stage. These programs foster farmer networks; assist with knowledge transfer and farmer-to-farmer training; unite peasant workers and farmers; and advocate for food sovereignty, reduced reliance on industrial methods, empowering agrarian citizens through training in agroecological policy and practices, increasing land access, and policies that support sustainable farming and enhance public awareness. These groups and organizations facilitate the kind of social cohesion in agrarian communities that is important for ensuring the long-term outlooks necessary for sustainable agriculture.

ECONOMIC SUSTAINABILITY AND FOOD PRODUCTION

In chapter 2, we noted agriculture is only a small part of our global economy in spite of the inescapable fact that food is a life necessity and forms the foundation of human well-being for all people. This seeming paradox arises from an agriculture that has transformed food into a commodity that is bought, sold, and traded like non-life-essential commodities.[20] The commodification of food means that developing countries are encouraged to produce food not just for their own needs but as an economic tool to gain footing in the global economy. Agriculture is often seen as a stepping stone in economic development, the ultimate goal being to transition from agriculture to industrial or service-based economies.

Agricultural Development

Agricultural development has proved useful in bringing less developed nations into the global economy, though, as we noted in chapter 1, it has often come at the price of environmental degradation and social dislocation. Industrial agriculture in particular, as opposed to the smallholder sustainable farming it often replaces, has served as a valuable first step in a developing nation's industrialization and attempts to lift itself out of poverty. Given that three-quarters of the poor people in the developing world live in rural agrarian communities where agriculture is the dominant livelihood, agricultural development is one form of development that could address rural poverty.[21]

While agricultural development makes sense as a first step in a developing nation's economic strategy to enter the global market, it is a tough path to take. The global market is a highly competitive arena dominated, not surprisingly, by industrialized countries whose trade laws, tariffs, subsidies, and other agreements keep market values for agricultural products low. Furthermore, agricultural development is often accompanied by land degradation and community reliance on external inputs like patented GMOs, synthetic fertilizers, and biocides. National food security can also decline dramatically if agricultural development focuses on nonfood crops like cotton, coffee, chocolate, sugarcane for ethanol, or oil palm for food additives and biodiesel, leading to a necessity to import food from other countries. Even if the focus is on food crops, industrial production invariably consists of monocultures, which still means a need to import food to ensure nutritional balance.

INTERNATIONAL INVESTMENTS AND INTERNATIONAL AID

Agricultural development most commonly starts with international investments (or capital flows) or international aid. International investment in a country's agriculture can boost its economy, but absolute gains are determined by capital returns to investors. This is especially common in the agricultural sector, where local, or in-country, agribusinesses are technically owned by a developing country (often by its upper economic classes) but in actuality function like local branches of transnational companies who command significant fractions of the financial gains.[22]

The alternative, in which foreign countries directly contribute capital toward agricultural development as a form of international aid, comes without the limits associated with foreign investments, but its effectiveness is dependent on the recipient country's capacity, economy, and governance (e.g., extent of transparency, democracy, and corruption) and on the form of the capital. The capital can be financial (e.g., currency), manufactured (e.g., equipment), or human (e.g., training in biotechnology).

Short-term aid, such as aid delivered during a famine, can be in the form of food, but its effectiveness is similarly dependent on a country's capacity, economy, and governance.[23]

CENTRALIZED MARKET CONTROLS

Global trade is not a new phenomenon, but it has been significantly deregulated over the last several decades, leaving much of global trade to be controlled by transnational corporations. For example, only four firms, Archer Daniels Midland (ADM), Bunge, Cargill, and Louis Dreyfus, currently hold 75 to 90 percent of the global trade in grains and oilseeds and control 60 percent of the trade, storage, and processing of these commodities. Just two transnational corporations, Cargill and ADM, control the export of 40 percent of the corn and 65 percent of the soybeans produced in the United States. Other examples include 83 percent of cocoa production being controlled by three companies, 85 percent of tea production controlled by three companies, and 75 percent of bananas controlled by five companies.[24]

The effects of such centralization of market controls are both positive and negative. Large corporations, for example, can make big advances in agriculture through financed research and development, and they have increased food production and availability in developing countries. On the other hand, their extensive domination over high-quality land in developing countries has made it impossible for small farmers not under their umbrella to enter the market.[25] Another example of a positive outcome is large-scale businesses generating and increasing access to markets to small-scale producers. Such positive outcomes, however, are often offset by the costs imposed by large-scale businesses for access to the markets they generate. For example, both Walmart and Amazon expanded their retail in organic produce, with the intent of making prices competitive with nonorganic equivalents. Organic food is generally more expensive because it requires more labor and care to produce. Walmart and Amazon, however, do not compensate for these added costs by lowering their retail markups. The organic farms ended up having new consumers who choose organic over industrial or conventional because of competitive pricing, but the farmers receive less for their produce. There are obvious environmental and consumer benefits to increasing a preference for organic products, but the profit gains were accrued by the transnational retailers, not the organic farms.[26]

TRADE POLICIES AND AGREEMENTS

Agricultural trade policies and agreements are established to protect farmers from booms in other countries and busts in their own countries, ostensibly keeping prices level. However, agricultural trade has experienced extreme growth in the last several decades, increasing by 4 percent in 1990–2002, which is twice as fast as agricultural production.[27]

Policies and multinational trade agreements in agriculture are governed predominantly by the World Trade Organization (WTO), an international governing body with 164 member countries as of July 2016. The WTO was established in 1995, stemming from the deliberations of the General Agreement on Tariffs and Trade (GATT), a series of trade negotiations that began after World War II. The Uruguay Round (1986–1994) was one of the most recent rounds of negotiations and led to the creation of the WTO.[28] In its initial incarnation, the GATT discussed tariff reductions so that developed countries could export excess food production. This was done with the intention of developing agricultural markets, though it was also used to export food aid to developing countries, leading to mixed outcomes for those countries.

One of the outcomes of the Uruguay Round was the Agreement on Agriculture (AoA), which had many stipulations for agricultural trade, but the main results were tariff reductions.[29] While tariffs were reduced less for developing countries, some argue that the percentage by which tariffs were reduced affected developing countries more than rich countries. Additionally, the final outcomes of the AoA did little to regulate export subsidies, which support the production of commodity crops. This allowed the United States, European Union, and other rich, industrial nations to continue what some call the *dumping* of commodity crops, wherein crops are exported in large quantities because national subsidies increase their production.[30] Inconsistencies such as these from the Uruguay Round led to some resentment of the AoA and triggered protests of future meetings (e.g., the WTO protests in Seattle in 1999). More recent adjustments to the AoA seek to regulate export subsidies and mitigate commodity dumping.[31]

To illustrate the complexities of trade agreements in agriculture, let's consider the North American Free Trade Agreement (NAFTA), one of the most controversial trade agreements established in recent history. NAFTA is an agreement between Canada, Mexico, and the United States that reduces import tariffs between the countries. The countries have separate agreements among themselves, and the most fraught and contested part of the agreement is between Mexico and the United States.

NAFTA, signed in 1994, ultimately eliminated tariffs, duties, and quantitative restrictions between the three countries. While this was supposed to happen progressively, the United States simultaneously increased export subsidies for maize by 3 billion dollars and drastically increased its exports to Mexico, flooding the market there. This depressed maize prices in Mexico, leading many maize farmers there into destitution and causing unrest in agricultural communities that was squelched by the Mexican government.[32]

In 2018, however, at the insistence of the United States, NAFTA was renegotiated and is proposed to be replaced by the United States–Mexico-Canada Agreement (USMCA). Opinions vary enormously as to how agricultural trade will change under the USMCA, if it is ratified by all three governments. Some argue there will be little impact; others argue the opposite.

The take-away message is that there are economic processes that have powerful local and global influences on agricultural production and markets. Often, however, trade policies and agreements are poorly connected to social or environmental factors, which hinders our ability to achieve agricultural sustainability in the near future.

SUMMARY

Sustainable food production is a function of both natural and social factors. Natural ecosystems consist of populations of plants, animals, and microorganisms that are highly dynamic, but over large spatial and temporal scales, their functioning and the services they provide are relatively stable—they exhibit long-term constancy and are often resistant and resilient to perturbations common to the region, like occasional droughts, fires, or storms. In contrast, agroecosystems are anthropic systems, so their stability, which is critical to sustainable production, is largely a matter of how people manage them. The majority of farmers, however, are poor and often lack the necessary capital to manage their small holdings in a way that ensures stability and sustainability. In contrast, wealthy farmers and agribusinesses, who own the majority of agricultural land, have greater management options and produce the large quantities of cheap food needed by the billions of urban and rural poor. Agricultural development, in principle, could generate revenue to reduce poverty and facilitate sustainable food production, but the highly competitive nature of world markets, agricultural policies, and global trade agreements among nations tends to favor developed nations and transnational agribusinesses. It is not surprising, therefore, that the poorest developing nations are those that are heavily reliant on agricultural development. These issues of system stability and inequalities are challenging but addressable. Research into environmental sustainability continually improves farming (e.g., climate-smart agriculture), while social sustainability is advanced through research in the social sciences (e.g., environmental-economic analyses of nonmarketed ecosystem services) and by social movements focused on equity, awareness, participation, and social cohesion within and among all sectors in agriculture.

4

FOOD, FARMING, AND HUMAN WELL-BEING

HUMAN DEVELOPMENT AND HUMAN WELL-BEING

The Three Pillars of Human Development

In chapter 1, we considered that development from an ecological perspective is a form of niche construction. Something members of almost every species do is modify their homes and habitats to better their lives. In the course of a few millennia, however, humans have vastly exceeded anything any other species has ever been able to do. So extensive is our niche construction that we have modified the entire planet to produce and sustain approximately 471 million tons of human mass: about 7.6 billion people. We noted, however, that while our biological success in terms of mass and numbers is indisputable, from ecological and social perspectives we've done poorly, creating a world in which environmental sustainability is at risk and prosperity is massively unequal. The relatively new paradigm of sustainable development has thus grown in popularity because it promises improved human prosperity while reversing, or at least preventing, further environmental harm that would adversely affect both future human generations and all other life on Earth.

Human endeavors to improve our lot in life are collectively known as HUMAN DEVELOPMENT, but there are many ideas about how best to define this term. Perhaps the best definition, and the one we adopt here, is that provided by the United Nations Development Program (UNDP). Their definition of human development is "the process of enlarging people's choices. In principle, these choices can be infinite and change over time. But at all levels of development, the three essential ones are for people to lead a long and healthy life, to acquire knowledge and to have access to resources needed for a decent standard of living. If these essential choices are not available, many other opportunities remain inaccessible."[1]

This definition or conceptualization of human development was pioneered by the economists Mahbub ul Haq and Amartya Sen and by the ambassador and UN senior advisor Üner Kirdar in the 1990s, and it highlights the importance of the noneconomic dimensions of development.[2] Embedded in this framework are what we can consider the three PILLARS OF HUMAN DEVELOP-MENT: health, education, and access. Here, *access* refers to access to information, water, and energy and other resources, natural or otherwise, including, for farmers, land, seed, livestock, fertilizers, biocides, and the other forms of capital described in chapter 1. We'll consider the important issue of access, which all three pillars of human development depend on, at the end of this chapter.

The Five Constituents of Human Well-Being

The defining objective of human development is to improve HUMAN WELL-BEING, but this term also lacks any universal definition. Scholars consider the concept of human well-being to consist of two distinct ideas: happiness and the actualization of human potentials.[3] Concrete definitions of human well-being, however, are rare,[4] with most individuals, fields, and institutions preferring to use descriptions of the dimensions or constituents of well-being, that is, what makes well-being *possible*, rather than trying to define to everyone's satisfaction what well-being *is*.

Focusing on the constituents of well-being, however, still requires us to choose among many different ideas of what those might be. To be consistent, we again resort to using the United Nations' set of constituents as it reflects one of global consensus. Human well-being is integral to the UN's Millennium Ecosystem Assessment's (MEA) framework,[5] which links biodiversity with ecosystems, ecosystems with ecosystem services, and ecosystem services with human well-being. However, the MEA's Ecosystems and Human Well-Being for Assessment[6] framework avoids defining human well-being directly. Rather, as many frameworks do, it describes well-being within the context of five constituents:

1. basic material for a good life (e.g., sufficient food, adequate shelter);
2. health (e.g., strength, wellness, longevity);
3. freedom of choice and opportunities (e.g., choices and opportunities for fulfilling valued goals in life);
4. security (e.g., safety, robust access to resources in the face of uncertainty); and
5. good social relations (e.g., social cohesion, respect, fair trade).

Figure 4.1 illustrates the MEA framework, with human well-being expanded into its five constituents, all of which are coupled to biodiversity and the ecosystems and services it supports. Note that freedom of choice and opportunity concern all four of the other constituents (figure 4.1, double arrows to right of figure).

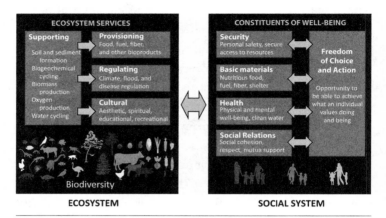

FIGURE 4.1 Biodiversity, ecosystem services, and human well-being.

The Millennium Ecosystem Assessment (MEA) framework (figure 2.1) links human well-being (right box) to ecosystem services (left box). The framework is valuable because it motivates us to consider that all ecosystem services are derived from the collective biological processes of the enormous diversity of plants, animals, and microorganisms that reside in ecosystems. We modify ecosystem services by changing biodiversity, reducing the abundance of, or even eliminating, those species we don't value and increasing the abundance of those we do. The framework is also valuable because it points out something we often forget or fail to recognize, which is that all the constituents of human well-being, including security, social relations, and even freedom itself, are linked to biodiversity through the services they provide.

Source: Authors.

Our framework for sustainable food and farming, illustrated in figure 1.8, is based in part on the MEA framework, but to keep things simple we just put a box around *human well-being*. We did the same in our framework for ecosystem services illustrated in figure 2.1. As well-being is at the heart of human development, however, in this chapter we unpack this box.

Human development and human well-being are enormous topics; thus, we limit our treatment of these topics to the focus of this primer: sustainable food and farming. Food security and the importance of social relations in food and farming appear in chapters 1, 2, and 3. In this chapter, we concern ourselves with the three other constituents of well-being:

- Food necessary for a good life (including essential and non-essential foods)
- The relationship between food and health
- The freedom to choose the food we produce, procure, and consume

Human Development Versus Sustainable Human Development

Note that human development, with its objective to improve human well-being, is not tied to any notions of sustainability, and indeed, since the beginning of our species some 200,000 years ago, notions of sustainability have played little to no role in our development and well-being. Slowly, however, global aspirations and concrete commitments to transition from traditional development to sustainable development have emerged, such as those outlined in the UN Sustainable Development Goals, which we considered briefly in chapter 1 and will consider in more detail in chapter 5. We will also explore in greater detail how the distinction between human development and SUSTAINABLE HUMAN DEVELOPMENT is seldom apparent in global deliberations concerning development more generally.

FOOD NECESSARY FOR A GOOD LIFE

Food

Recall from chapter 1 that, except for plants, which can feed themselves or obtain the energy they need from light, and certain microorganisms that obtain energy from chemical sources, the majority of life on Earth feeds on other organisms to get their energy. The distinction between self-feeders, or *autotrophs*, and those that feed on others, or *heterotrophs*, is all about where organisms get their energy. Feeding is not just about energy, however; food is a mix of energy and nutrients. It is about getting the nutrients one needs to build the four classes of biomolecules all living organisms are composed of: carbohydrates, proteins, lipids, and nucleic acids (chapter 1). The terms *autotroph* and *heterotroph*, although useful ecological constructs, are misleading, since feeding isn't just about energy capture. A bottle or package of plant food (fertilizer) doesn't list its energy content but instead tells you about the minerals found in the powder or solution. When it comes to animals, however, whether wildlife, livestock, pets, or ourselves, the energy content of food is often considered the most important factor. We know, for example, that food contains a mix of carbohydrates, fats, and proteins, which is why they are listed on the nutrition-fact labels of packaged foods (figure 4.2), but there are many other components of food that are important, such as vitamins, minerals, essential fatty acids, and essential amino acids—too many to list on a label. We will discuss nutrients in detail later on, but for now, we simply note that food is complex and that a diet that contains all the energy and nutrients one needs generally requires a diversity of foods.

Interestingly, nutrition-fact labels are not put on things like fruits, vegetables, meats, and dairy products found in farmers'

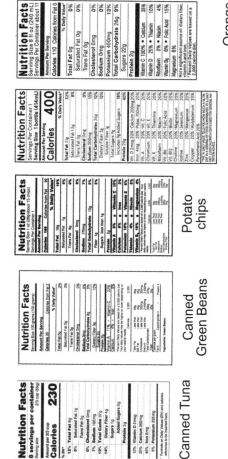

FIGURE 4.2 Examples of nutrition–fact labels on packaged foods.

Note the prominence of carbohydrates, fats, and protein.

Source: Authors.

markets, greengrocers, bakeries, and butchers. It's only when food is processed, preserved, or otherwise treated and packaged that it is labeled with nutrition facts.

While it might seem strange to ask the following question, it is instructive to do so: Why is it that the other 8.7 million species on Earth that feed, grow, and reproduce successfully every day don't need nutrition-fact labels like we do (if they could read)? All species make food choices, to be sure. Predators feed on animal prey, so their diets are rich in protein, while herbivores feed on plants, so their diets are rich in carbohydrates. Omnivores feed on both plants and animals, so their diets are varied. No matter what their dietary lifestyles are, however, species other than humans never need their foods labeled.

When it comes to organisms in nature, their behaviors, digestive physiology, and diet choices reflect evolutionary and ecological outcomes that allow them to survive in their habitats. For many species, especially vertebrates such as birds and mammals, diet choice is partly learned by offspring from their parents, and many animals will experiment and try out novel foods; however, by and large, animals find the foods they need in the diversity of life around them. This does not mean that food is abundant and nutritious—indeed, most organisms struggle to stay alive, with the majority of their waking life spent foraging, feeding, and digesting. Ours is a planet typified by life in perpetual hunger.

Until 10,000 years ago, before the advent of farming, our lives as omnivores were not much different from those of our fellow species. The diversity of life in our habitats and our hunting and gathering behaviors were well matched with our digestive physiologies. Like all animals, our preagricultural ancestors undoubtedly went hungry at times and suffered dietary deficiencies when certain foods were in short supply. In the modern world, the diets of humans are more a matter of what foods are grown, are available for consumption, and are accessible to those who want or need them, and they are subject to choices we make when it comes to selecting what we will and will not eat. Except for a few remaining tribes of hunter-gatherers (e.g., the Kalahari Bush People, Sentinelese of the Andaman Islands, or Pirahā of the Amazon), food for the majority of the world's people consists of what we farm and/or obtain from markets. For some people, the modern-day dietary landscape has become immensely complex. Once we did not need labels; we just needed a habitat rich in diversity and to learn from our parents what to eat. Now, we need food labeled.

When food items are processed and packaged, they may not be readily associated with the plant and animal sources of their origins unless labeled with nutrition facts or pictures of the grains, legumes, fruits, vegetables, animals, or farms from which they came. To the uninformed, flour is not recognizable as wheat, jelly is not recognized as fruit, the contents of a can of tuna are not recognized as fish, and potato chips are not recognized as potatoes. Few may know what marshmallows are (sugar and gelatin) or that the marshmallows in Lucky Charms consist of sugar, modified corn starch, corn syrup, dextrose, gelatin, calcium carbonate, yellows 5 and 6, blue 1, red 40, and artificial flavor. Similarly, until the Beech-Nut Corporation conceded in 1987 that much of its apple juice for babies contained no apple juice, parents who purchased the product would not have known that their babies were being fed water, sugar, and a variety of additives—but nothing from apples.

In the United States, it wasn't until 1973, in the face of a rapidly increasing and bewildering array of packaged and processed foods, that the Food and Drug Administration required packaged foods to carry nutrition-fact labels to help consumers make informed choices with respect to meeting their dietary needs. And it wasn't until 1994 that the more detailed nutrition-fact labels, such as those in figure 4.2, came into force.

As a final example, the McDonald's fast food chain's immensely popular Chicken McNuggets do, in fact, contain chicken (~50 percent), in the form of a boneless, homogenized chicken meat paste, but the other 50 percent consists of water, canola oil, corn oil, hydrogenated soybean oil, bleached wheat flour, niacin, reduced iron, thiamine mononitrate, riboflavin, folic acid, yellow corn flour, vegetable starch (modified starch from corn, wheat, rice, and peas), salt, baking soda, sodium aluminum phosphate, sodium acid pyrophosphate, calcium lactate, monocalcium phosphate, spices, yeast extract, lemon juice solids, and dextrose. Many of these ingredients are, in fact, valuable nutrients (e.g., riboflavin, niacin, reduced iron), but, in this case, even a knowledgeable consumer may not understand what all these ingredients are, how they are produced, or the effects they may have on the body. McDonald's is also quick to point out that there are no artificial flavors, colors, or preservatives in Chicken McNuggets and that the Olympian runner Usain Bolt ate one thousand McNuggets as part of his dietary regimen at the Beijing Olympics in 2008, in which he won three gold medals. A consumer is likely to assume that because the product's dominant ingredient is chicken and because the company appears to care enough about health to not use artificial flavors or colors—and that if an Olympic gold medalist eats them (along with millions of people eating them daily)—it is okay to consider these part of one's diet. They are also inexpensive and filling, which for many may be the decisive factors influencing the choice to consume this product.

A Good Life

Note that human well-being, as we have defined it here, includes the rather enigmatic idea of being able to lead a *good life*. What constitutes a *good life*, however, will vary from culture to culture and from person to person. From our deliberations about development and well-being, health, longevity, leading a life that meets what we define as a decent standard of living, and the freedom to make choices would almost assuredly be part of anyone's idea of a good life. Food, we argue, is going to be the dominant factor in these determinants of a good life. Without sufficient, nutritious food, leading a good life is just about impossible.

While a *good life* means different things to different people, we can begin to explore the issue by considering what the dietary minimums would be for a poor life. What sort of diet would be one that is sufficient to keep one just barely alive? Diets above such minimums would be the starting point for a good life. Such dietary minimums reflect meeting basic metabolic needs sufficient to stay biologically alive for at least one day. While such a life would be far from good, it at least might come with the hope that tomorrow may be better.

As food consists of energy and its nutrients, identifying the minimums for a good life necessarily focuses on these. *Nutrition* is the term that we use to capture both energy and nutrients, and a *nutritious* diet is one that is a sufficient mix of the two such that an individual can grow, survive, and reproduce. What constitutes a nutritious diet for an individual depends on a number of factors, such as an individual's age, height, weight, level of physical activity, and, for women, whether they are pregnant or breastfeeding. To simplify things, we will primarily consider the diet and nutrition for an average adult person.

It is important to emphasize that a life of minimums is far from a good life, particularly when it comes to nonessential foods and other agricultural products that are nevertheless important to well-being, such as coffee, tea, spices, flavorings like vanilla, alcoholic beverages, natural pharmaceuticals, and perhaps even some things that might be seen as harmful but are still desired, like tobacco or betel nut. These are all nonessential for dietary health but may be deemed essential for a "good life." For the next sections, however, we focus on nutrition, health, and safety.

Energy

Irrespective of individual differences, everyone needs energy to carry out metabolic processes, such as respiration, digestion, the production and processing of biological waste, development, reproduction, and immunity (defending one's body from infectious diseases and parasites), and energy is central to brain function and being physically active. Given its importance, it's worth taking a closer look at dietary energy.

First, how is dietary energy measured? The energy content of food is typically measured in kilocalories (kcals) (or thousands of calories, though this is often reported simply as *calories*, which can create some confusion). A calorie is often misconstrued as a measure of weight, fat, or sugar content, but it is actually a term from physics. One calorie is equivalent to 4.18 joules. A joule is a force of one newton applied on an object over a distance of one meter, and a newton is the force it takes to accelerate 1 kg from a stationary point in a fixed direction for one meter per second squared. If you are a physicist, such an explanation of a calorie might be useful, but for the layperson thinking about energy in food, it's easier to think of a calorie as a unit of energy with which one can do a specified amount of metabolic work (e.g., respiration, digestion, processing wastes, and so forth).

To better visualize a calorie of energy in food, consider that a teaspoon of sugar contains 16 kcals, which is enough energy to keep your metabolic functions going so you can stay alive for about 20 minutes, assuming you are at rest. We often think of this process of food energy driving metabolism as a body *burning* energy, much the way a car burns gasoline.

Further comparisons to help visualize calories can be made using the labels shown in figure 4.2 and the values listed in table 4.1, which provide some estimates of the number of kcals per gram of several common foods.

Ignoring differences in factors such as age, weight, height, and gender, we can now ask how much energy an average, active adult weighing around 140 lbs (64 kg) needs per day. The UN Food and Agriculture Organization puts it at between 2,400 and 3,700 kcals of energy, depending on activity levels. To put this in perspective, that's about 3 to 5 cups of sugar per day. The minimum caloric intake for a person at rest is half that amount. If your caloric intake is around 1,200 to 1,500 kcals, you will just about make it through the day, assuming you are resting. After an extended period of time below a daily intake of 1,000 kcals, you are likely to reach a state of extreme hunger, and your body will begin to draw on fat reserves as a source of energy; once your fat is gone, you will start to lose muscle tissue, a condition called *wasting*.

The FAO considers a daily intake of 2,500 kcals to be "well nourished," and we use this amount to mark what we consider as enough food for a good life. A daily intake of 1,480 kcals is considered undernourished, which we take to be the lower limit for leading any sort of life at all.[7]

TABLE 4.1 A SAMPLING OF NUTRITIONAL CONTENT (MACRONUTRIENTS AND MICRONUTRIENTS) PER GRAM OF FOOD

	Energy (kcal)	Protein (g)	Carbohydrate (g)	Total Fat (g)	Calcium (mg)	Iron (mg)	Sodium (mg)	Potassium (mg)	Phosphorus (mg)
Grains									
Wheat	3.41	0.14	0.73	0.00	0.35	0.04	0.05	4.08	3.48
Rice	1.31	0.02	0.29	0.00	0.10	0.00	0.01	0.35	0.43
Corn	3.66	0.08	0.78	0.01	0.05	0.01	0.03	1.62	0.84
Processed food									
Cap'n Crunch	4.00	0.02	0.37	0.02	0.00	0.06	3.38	0.48	0.35
Potato chips, plain	5.35	0.07	0.49	0.35	0.21	0.02	5.33	13.28	1.51
Chocolate-chip cookie	4.90	0.02	0.21	0.08	0.11	0.01	0.94	0.48	0.37
Vegetables									
Broccoli	0.35	0.03	0.10	0.00	0.52	0.01	0.54	3.83	0.87
Carrots	0.41	0.02	0.10	0.00	0.32	0.00	0.67	3.10	0.33
Mushrooms**	0.22	0.03	0.03	0.00	0.03	0.00	0.03	2.70	0.73
Tomatoes	0.18	0.02	0.08	0.00	0.19	0.00	0.10	4.63	0.48
Fruit									
Apple	0.52	0.00	0.30	0.00	0.13	0.00	0.02	2.35	0.24

Orange	0.47	0.02	0.24	0.00	0.83	0.00	0.00	3.76	0.29
Mango	0.64	0.02	0.29	0.00	0.16	0.00	0.03	2.56	0.17
Strawberries	0.32	0.02	0.10	0.00	0.21	0.01	0.02	2.05	0.32
Papaya	0.39	0.00	0.11	0.00	0.29	0.00	0.03	3.02	0.06
Dairy									
Milk	0.60	0.13	0.19	0.13	4.62	0.00	1.63	5.86	3.73
Yogurt	0.63	0.16	0.21	0.05	5.27	0.00	2.02	6.73	4.14
Cheese	4.04	0.19	0.02	0.27	5.73	0.00	4.94	0.78	4.06
Egg	0.60	0.13	0.19	0.13	4.62	0.00	1.63	5.86	3.73
Fish									
Cod	1.05	0.27	0.00	0.02	0.17	0.01	0.94	2.90	1.65
Salmon	2.07	0.27	0.00	0.14	0.17	0.00	0.73	4.57	3.00
Meat									
Beef	3.24	0.35	0.00	0.27	0.03	1.24	4.78	0.33	2.83
Pork	2.99	0.29	0.00	0.25	0.01	1.08	4.37	0.30	2.60
Chicken	2.39	0.32	0.00	0.16	0.01	0.98	2.65	0.27	2.17

* Units are per gram of food; for example, energy is in kcals per gram of food.
** Mushrooms are not vegetables but are grouped here for convenience.

Nutrients

The nutrient content of food is much more complicated because it involves a wide array of compounds that are important to an enormous variety of metabolic functions. To bring some order to the bewildering diversity of nutrients in our diets, nutritionists have divided nutrients into two main classes. The first class is the *macronutrients*, which are the familiar proteins, fats, and carbohydrates that are the sources of energy and building materials for growth, development, and reproduction. The second class is the *micronutrients*, which include essential minerals, essential fatty acids, essential amino acids, and vitamins. Micronutrients may not be needed in large quantities, but they are extraordinarily important for virtually all metabolic processes, from hormone and protein synthesis to serving key roles in enzyme function. Iron, for example, is important in hemoglobin's ability to transport

BOX 4.1A   DIETARY NUTRIENTS, PART 1

MACRONUTRIENTS

Three of the four classes of life's essential biomolecules are necessary in bulk quantities, which is why they are often listed on nutrition-fact labels:

- Carbohydrates
- Fats
- Proteins

All organisms, including humans, synthesize nucleic acids, so they are not necessary components of a healthy diet.

MICRONUTRIENTS

Small quantities of organic and inorganic compounds, collectively referred to as *micronutrients*, are needed for health and longevity. These are:

Minerals: Any element outside of carbon, hydrogen, nitrogen, and oxygen essential to the formation and functioning of key biomolecules in our bodies is considered a mineral nutrient. These are often divided into

- Macrominerals, which include *sodium*, *chloride*, *potassium*, and *calcium* (fluid balance, muscle, neuron function); *phosphorus* (bones, nucleic acids, and energy transport in cellular metabolism), *magnesium* (protein synthesis, muscle, nerve, and immune system functioning); and *sulfur* (numerous protein functions)
- Microminerals (or trace minerals), which include *iron* (blood, energy metabolism), *zinc* (protein, reproductive biology, growth, immune system and many other functions), *iodine* (thyroid hormone functions), *selenium* (antioxidant), *copper* (enzyme functions), *manganese* (enzyme functions), *fluoride* (bones and teeth formation), *chromium* (blood sugar regulation), *molybdenum* (enzyme functions), *nickel* (blood and bone function), *silicon* (bone, skin, and hair health), *vanadium* (under study, but implicated in many aspects of health), and *cobalt* (vitamin B12 function)

Essential fatty acids: We can synthesize most of the fatty acids we need, but there are two we cannot but that are important parts of a balanced diet:

- α-Linolenic acid (or omega-3 fatty acid)
- Linoleic acid (or omega-6 fatty acid)

BOX 4.1B   DIETARY NUTRIENTS, PART 2

MICRONUTRIENTS (CONTINUED)

Essential amino acids: Protein is the most diverse class of biomolecule in form and function, but they are made up of just 20 kinds of amino acids. Healthy humans can synthesize most of the amino acids they need, but there are nine that cannot be synthesized and must be obtained from food. These essential amino acids are histidine, isoleucine, leucine, lysine, methionine, phenylalanine, threonine, tryptophan, and valine.

Vitamins: There are thirteen organic compounds we cannot synthesize known as vitamins, all of which are essential:

- A (cell growth, vision, embryo development, reproductive function)
- C (numerous functions, including iron metabolism, hair and nail protein synthesis, and nerve function)
- D (can be synthesized in sunlight but is often deficient in many individuals; regulates blood calcium levels, which is important for proper bone formation)
- E (antioxidant; antioxidants help prevent harmful oxidation [the binding of oxygen] of compounds that can lead to the disruption of metabolic functions and key cellular functions)
- K (bone formation, calcium transport, blood clotting)
- B vitamins, or the vitamin B complex (initially thought to be a single vitamin, subsequent research revealed it to be a complex of related vitamins); eight water-soluble vitamins that perform a variety of functions

and are often coinvolved with one another's functioning.

- B1, or thiamine (carbohydrate, fat, and protein metabolism, nerve function)
- B2, or riboflavin (amino acid and fat metabolism, B6 and folic acid function)
- B3, or niacin (carbohydrate and alcohol metabolism, cholesterol regulation)
- B5, or pantothenic acid (energy metabolism and neurotransmitter, cholesterol, and hormone synthesis)
- B6 (amino acid and hormone metabolism)
- B7, or biotin (protein, fat, and carbohydrate metabolism)
- B9, or folate (nucleic acid synthesis)
- B12 (DNA replication, nerve function, and indirectly in heart, bone, and brain function)

oxygen, and antioxidants, such as vitamin C and selenium, can help prevent the dangerous, unintended oxidation of compounds that can disrupt a variety of cell functions and even lead to cancer.

Nutrients are briefly described in boxes 4.1a and 4.1b to provide some sense of the extraordinary diversity of roles they play in health and well-being.

FOOD FOR HEALTH

Health is the outcome of many factors, nutrition being one of the more well-known sources of poor health. A nutritious diet is one that provides all the energy and nutrients an individual needs to maintain a good life.

The World Health Organization (WHO) suggests that the ideal, healthy diet of a typical adult consists of the following:

- At least 400 g of fruit and vegetables per day, excluding potatoes, sweet potatoes, cassava, and other starchy roots.

- Less than 5 percent of total energy intake from sugars, which is equivalent to 25 g (or about 6 level teaspoons) for a person of healthy body weight consuming about 2,000 kcals per day. This includes sugars added to foods or drinks by the manufacturer, cook, or consumer, as well as sugars naturally present in fruits (and their juices and concentrates), honey, and syrups (e.g., maple syrup).

- Less than 10 percent of total energy intake from fats. Unsaturated fats (found in fish, avocado, and many nut and seed oils) are preferable to saturated fats (found in fatty meat, butter, palm and coconut oil, cream, cheese, ghee, and lard) and trans-fats of all kinds, including both industrially produced trans-fats (found in baked and fried foods and in prepackaged snacks and foods, such as frozen pizza, pies, cookies, biscuits, wafers, and cooking oils and spreads) and ruminant trans-fats (found in meat and dairy foods from ruminant animals, such as cows, sheep, goats, and camels).

- Less than 5 g of iodized salt (equivalent to about one teaspoon) per day.[8]

In chapter 1, we estimated what this means on a global level if 7.6 billion people each were to have such an ideal diet. It translated into staggeringly massive numbers; millions of tons of food have to be produced per day.

Nutritional Diversity

The macro- and micronutritional contents of individual foods vary enormously across classes, such as grains, vegetables, fruits, dairy, and meats, so a healthy diet, such as that recommended by the WHO, should includes food from across the different classes. Consider, for example, the basic nutrients of the foods listed in table 4.1 and how a diet made up of only one or two of these foods will lead to deficiencies or harmful excesses in one or more classes of nutrients.

Malnourishment

What if food, and thus nutritional diversity, is low and diets are significantly less balanced than what the WHO recommends? Globally, humans receive 75 percent of their calories from only 12 plants and five animals, and only three plants (corn, wheat, and rice) make up 60 percent of the plant species we eat.[9] This can happen in an individual's diet because the diversity of foods is low in one's habitat, farmers grow only one or a few crops, markets offer little in the way of options for a diverse diet, dietary choices are constrained by cultural and/or political reasons, a diverse diet is too expensive, or a combination of these factors. An imbalanced diet, driven by such factors, may lead to MAL-NUTRITION, which in turn can lead to numerous illnesses and diseases; dangerously low weight or, ironically, obesity; impaired mental development; reduced cognitive function; decreased physical or mental productivity; detrimental changes in hormone levels; stress; stunted growth; muscle wasting; and even death. Clearly, if one is severely malnourished, none of the three pillars or objectives of human well-being can be met.

The WHO identifies three categories of malnutrition: (1) UNDERNUTRITION, (2) micronutrient deficiencies or excesses, and (3) OVERNUTRITION. Undernutrition is associated with insufficient intake of energy, protein, or both and leads to wasting and stunting (permanently impaired growth).

Micronutrient-related malnutrition manifests in various ways depending on the particular nutrient, its function in the body, and whether intake is insufficient or in excess. Globally, the main micronutrient deficiencies are iodine, iron, vitamin A, and zinc. An iodine deficiency is associated with brain damage, iron deficiency can lead to anemia, a deficiency in vitamin A is associated with a weakened immune system, and zinc deficiency compromises immune system function and is associated with gastrointestinal infections. Nutritional edema (swelling) can also occur in the face of vitamin and mineral deficiencies.

Overnutrition occurs when diets that are nutrient poor and energy dense (i.e., high in fat and added sugar) contribute to an imbalance where energy intake exceeds energy expenditure. Such imbalances, especially when individuals lead physically inactive lives (e.g., no manual labor or exercise), lead to an individual becoming overweight and, if unchecked, develop into obesity. Diseases related to being overweight or obese include type 2 diabetes, liver disease (cirrhosis and nonalcoholic fatty liver disease), cardiovascular disease (e.g., high blood pressure, high cholesterol, heart disease), kidney disease, hypothyroidism, some types of cancer (e.g., colon, breast, gallbladder, liver), osteoporosis, anemia, and dental disease (e.g., cavities).

Dietary diseases caused by malnutrition range enormously in their symptoms. They may

- affect metabolic processes by altering the function of, release of, and the body's response to hormones;

- reduce circulation;
- impede filtration and removal of wastes;
- lower immunity;
- affect mobility;
- cause edemas;
- contribute to problems with eyesight;
- decrease organ or system functioning;
- contribute to respiratory problems;
- result in weakness and lethargy; and
- result in poor hair, skin, and nails.

If the dietary pattern (and lifestyle) contributing to the development of a given disease does not change, individuals may become dependent on medicines to control their symptoms, take more drastic measures such as surgery to remove or repair damaged tissue, or find themselves with a disease that has progressed to the point that it is no longer reversible. Each of these outcomes affects an individual's physical and emotional well-being. These scenarios also incur monetary costs, which may or may not be manageable depending on an individual's socioeconomic status. In situations of deprivation and/or lack of health insurance, individuals may find themselves choosing between, for example, treating their (or a family member's) disease, feeding family members, or feeding themselves.

The Burden of Malnutrition

In a world that touts global food sufficiency, the prevalence of malnutrition is shockingly high. The causes for malnutrition are many and varied and include biological (e.g., a rise in food allergies), social and demographic (e.g., poverty), environmental (e.g., adverse weather and climate-change impacts on food supplies), and behavioral factors (e.g., lifestyles becoming less active in urban settings).[10]

Children are particularly sensitive to the adverse consequences of malnutrition. While undernutrition in children has declined, the most recent joint report on childhood malnutrition by the United Nations Children's Fund, WHO, and World Bank Group estimates that nearly half of the deaths of children under five are attributable to undernutrition.[11]

In contrast to undernourishment, being overweight or obese in both children and adults is on the rise. From 1975 to 2016, obesity has tripled globally, with 39 percent of adults classed as overweight and 13 percent obese in 2016 and 16 percent of children and adolescents aged 5 to 19 overweight or obese.[12] In 2017, nearly 6 percent of children under the age of 5 were overweight. While the prevalence of obesity appears to have plateaued in high-income countries, the increasing trend in obesity continues in low- and middle-income countries. Regions in Asia and Africa experienced the greatest increases in child and adolescent (ages 5–19) obesity. Currently, Africa and Asia bear the greatest burden of malnourished children in all its forms.

Overall, the burden of diet-related disabilities in the 2017 Global Burden of Disease report,[13] estimated as years lost to disability (YLD), is astonishing, given that such a burden is entirely preventable if food security could be made a universal right (figure 4.3).

While human undernutrition is still a problem globally—and, in fact, the number of hungry people went *up* in 2016—diet-related diseases (such as diabetes and heart disease) are now a leading cause of death in the industrial world and the number-one cause in the United States.[14] Though we produce more food than ever before, it is increasingly apparent that this food is providing

Dietary Risks (Both sexes, all ages, 2017, YLD per 100,000

FIGURE 4.3 Number of years lost because of disabilities (YLD) caused by dietary-related diseases.

When quantifying the impact of adverse health conditions, one metric that is useful years lost because of disability (YLD), or years lived in poor health. YLD is measured by taking the prevalence of the condition in the population (in this figure, the population of a nation) multiplied by the severity of the disease, which is determined through surveys of the general public. In every population, there are multiple health risks each individual faces. This figure focuses on health risks associated with poor diets. For example, in this figure, if a nation has a YLD of 1,000, it means that for every 100,000 years of life all its people have lived, 1,000 years were lost to dietary disease. Note that the burden of dietary diseases can be quite high for developed nations.

Source: Data for this figure were derived from https://vizhub.healthdata.org/gbd-compare/.

inadequate nutrition; this is a function both of the food available (access) and consumers' dietary choices.

These outcomes highlight the need for education on the links between diet and disease and early interventions that can prevent or reverse diet-related diseases. As indicated by the discussion on malnutrition, education throughout the life cycle is crucial: diet-related diseases are prevalent in children, adolescent, and adult populations. Interventions are also needed across the life cycle. Further, interventions should be implemented at the level of the individual, community, municipality, and country to address the prevention and treatment of diet-related diseases from multiple angles (e.g., via individual nutrition, food security, social protection measures, development goals, and food policy).

Food Safety

Access to sufficient, safe, nutritious food is a human right and a component of food security. Food safety matters at two levels. There are safety issues associated with the production of food, such as the appropriate use and handling of equipment, the safe application of pesticides and herbicides, appropriate protective clothing, reasonable working hours, and injury prevention and care. With respect to eating, food safety is concerned with both preventing the contamination of food and preventing the ingestion of food that may be contaminated with pathogens and/or chemicals. Food may be contaminated throughout the supply chain, from producer to consumer. Thus, safety measures that reduce the risk of contamination should be in place throughout the supply chain, and both food service providers (e.g., restaurants) and consumers should take it upon themselves to handle,

prepare, and store food safely to reduce the risk of spreading and/
or contracting foodborne illnesses.

FOODBORNE ILLNESSES

Foodborne illnesses (also referred to as "food poisoning") are
caused by pathogenic bacteria, parasites, viruses, and molds
that, once inside an individual's body, cause distress. Often the
distress is gastrointestinal; however, depending on the pathogen, other symptoms may be present, including fever, head-
ache, neurotoxicity, and loss of motor function. According to
the Center for Disease Control, the top five pathogens that
cause foodborne illnesses in the United States are Norovirus
(virus), *Salmonella* (bacteria), *Clostridium perfringens*, *Campylo-
bacter* (bacteria), and *Staphylococcus aureus* (bacteria). However,
the pathogens *Escherichia coli* (bacteria), *Shigella* (bacteria), and
Hepatitis A (virus), along with Norovirus and *Salmonella*, are
referred to as the "Big 5" pathogens because they are highly
infectious, require a low dose for infection, are easily transmit-
table, and can cause severe symptoms. These pathogens are also
implicated in causing foodborne illnesses across the globe.
Children under the age of five are at the greatest risk of becoming sick from contaminated food.[15]

CHEMICAL CONTAMINATION

Chemical contamination may result in acute cases of poisoning
or long-term disease such as cancer. Chemical contamination of
food may occur through the application of pesticides and herbicides, using contaminated water to irrigate crops, or growing
crops in contaminated soils or soils high in elements that may
be detrimental to human health if consumed above a certain
amount (e.g., heavy metals and selenium). Individuals may also
experience chemical poisoning by ingesting foods such as corn

and mushrooms (both sources of mycotoxins) and marine organisms (a source of marine biotoxins). Another chemical concern is the ingestion of persistent organic pollutants (POPs). These are organic compounds that are resistant to biodegradation and thus persist in the environment for a long time, can be transported long distances, and can both bioaccumulate (build up in an organism) and biomagnify (the buildup of a chemical from one trophic level to the next as an organism consumes another, thereby concentrating the chemical in organisms at the top of the food chain). Exposure to POPs may increase an individual's risk of cancer and reproductive disorders, increase the risk of birth defects, and serve as endocrine disrupters.

Preventing both the contamination of food and the ingestion of contaminated food are crucial to stopping the spread of foodborne illnesses. Prevention includes policies that enable safe food systems, including transparency and collaboration (i.e., cross-border) in supply chains; education and training in safe food harvesting, storage, and transport for producers, distributors, and other food handlers (e.g., farmers' market vendors and food service workers); and, for food preparers (e.g., chefs) and consumers, adhering to the four basic practices of food safety: clean, separate, cook, and chill. Access to sanitation and to safe and clean water is important in preventing both contamination and the spread of illness, including foodborne illness. However, in many regions of the world, communities do not have access to sanitation facilities or safe and clean water. There is a global effort to provide universal, affordable, sustainable access to water, sanitation, and hygiene (WASH). Access to WASH is goal 6 of the 2030 AGENDA FOR SUSTAINABLE DEVELOPMENT,[16] the realization of which could improve food safety, among other positive outcomes, such as contributing to a reduction in malnourishment, reducing WASH-related child mortality, and reducing the risk of young women being subject to attack while collecting water.

FOOD AND FREEDOM OF CHOICE

Food is a basic human need, something everyone needs before they can do anything else with their day, week, or life. Without food security, choices are limited; it's a priority for survival. While in some places people may have an overabundance of food options, in many parts of the world food choices are limited to what someone can grow, what's available at a local market, or accessible food aid. Different forms of food access provide varying forms of freedom in choice at the levels of both food production and food consumption.

Dietary Preferences and Restrictions

Dietary needs are dictated by biology, but what an individual eats is determined by a variety of factors that give rise to food preferences (i.e., what an individual desires and chooses to eat) and dietary restrictions (i.e., what an individual will not or cannot eat). Some cultural dietary prescriptions can be salutary, such as vegetarianism, or harmful, such as a preference for red meat and disdain for vegetables.

Food preferences and restrictions are often informed by social factors such as religion, socioeconomic status, culture, education, ethics, morals, and values. Various religions restrict the consumption of certain foods on a daily basis and/or restrict the consumption of any food (i.e., fasting) in observance of holidays. Religious practices may also require certain foods to be consumed in observance of holidays or life events. Culture informs food preferences that are often linked with geography (i.e., cuisine type) and a sense of place or that are reflective of diet- and food-related trends. In this digital age, there is also no shortage of diet and food information and misinformation that can shape food trends and food preferences for those with internet access. For many, nonreligious beliefs also influence preferences, such as subscribing to the idea that it is unethical to eat animals or animal products, believing that locally produced food is superior to imported foods, a holding a preference for organically or industrially produced foods, disapproving of GMOs, subscribing to the idea that only fair-trade or environmentally friendly certified foods should be purchased, or voicing concern over wildlife protection and thus not eating foods containing oil palm products that lead to habitat destruction and threaten endangered species like orangutans.

As another example, over the past few decades the UN has pushed for the cultivation and consumption of insects to replace large animal livestock as a protein source. Insects can be an excellent protein source and can be produced using minimal feed and very little land; they are thus a sustainable alternative to meat. Though insects are traditionally incorporated in cuisines around the world, they are not a staple in the Western diet. A 2013 UN report notes that people not accustomed to eating insects see them as dirty and have difficulty getting over the "ick" factor. Though a more sustainable option than livestock, the majority of Western consumers nevertheless deem insects culturally inappropriate for consumption.[17]

Dietary preferences and restrictions add to the complexity of human development as well as complicate progress toward achieving environmental sustainability. Steps in food production that may improve the balance between inputs and outputs and improve intergenerational equity supported by ecological principles (e.g., vegetarianism as a means to improve overall energy efficiency in food production) or technological innovation (e.g., GMOs) may be not be options in the face of dietary restrictions. Human development strives to improve choices and freedoms, and dietary preferences and restrictions reflect freedoms and choices. Indeed, as described in chapter 1, ensuring the rights of individuals and peoples to adhere to their preferences and choices is integral to food justice and sovereignty.

The Contingency of Access to Exercise Choice

A key element to food security, food sovereignty, and food justice is access to sufficient, safe, nutritious food. Access, for example, is considered by the FAO to be a critical component of food security. To be food secure, an individual must have physical, social, and economic access to food that meets their dietary needs and preferences and that supports an active and healthy life (i.e., their well-being).[18] In chapter 1 we noted eight factors central to food security, and we repeat them here: food should be (1) available, (2) adequately accessible, (3) reliable, (4) in sufficient supply, (5) culturally appropriate, (6) safe, (7) nutritious, and (8) capable of sustaining a healthy and active life.

Physical access concerns the location of food ready to be consumed and/or purchased, that is, where food is distributed and whether an individual can access that location. Physical access is determined by factors such as infrastructure (roads and buildings); policies that affect trade and distribution; conflict; transportation availability and routes (for both distributors and consumers); weather and natural disasters; market or store opening and closing times; and consumers' obligations, schedules, and any functional limitations (e.g., disability or injury). Areas that lack physical access to affordable, healthy food (e.g., fruit, vegetables, whole grains) are called *food deserts*. Food deserts are present in both rural and urban areas: anywhere there is a lack of providers (e.g., grocery stores, farmers' markets) selling healthy food and where traveling to acquire healthy food is not manageable. Food deserts are often associated with low-income or poverty-stricken areas as well as the double burden of malnutrition and disease caused by a lack of nutritious food and a reliance on processed foods sold at small convenience markets.[19]

Social access means an individual can access the food they need in a socially acceptable manner. This is generally taken to mean that an individual does not have to rely on charitable assistance, credit, government programs, stealing, scavenging, or begging to acquire food. On the other hand, food-related social protection measures may be beneficial in promoting access to food in areas where social structures, norms, or traditions reinforce inequalities (e.g., based on gender or race) or promote social exclusion. Economic access concerns an individual's or household's ability to purchase and/or grow food. Economic access is often discussed within the contexts of income, employment, and poverty (i.e., the inability to afford nutritious food), but households that rely on SUBSISTENCE FARMING may also be constrained by a lack of resources such as land, equipment, seed, livestock, capital, and/or labor. Household structure may also constrain economic access to food depending on the household's income and food requirements (e.g., the number of mouths to feed). In addition to socioeconomic demand-side (consumer) drivers of economic access to food, there are supply-side drivers that influence whether adequate amounts of safe, nutritious food can be accessed economically. For example, trade agreements, reliance on imports, market volatility, and other factors can affect the prices of food and other agricultural commodities (e.g., seed and fertilizer) and thus consumers' economic access to food.

Individuals or households may lack access to safe, nutritious food in one or more of these dimensions. The extent to which (e.g., frequency and duration) and the number of dimensions across which (i.e., physical, social, and economic) the lack of access is experienced will affect an individual's or household's level of food security (or insecurity). The consequences of decreased food security affect human well-being and the capability to lead a good life; these consequences could be short term and (relatively) easy to recover from or longer term, with lasting, negative impacts on human development.

SUMMARY

At the most basic, biological level, the trajectory of a human life concerns little more than ensuring that each individual lives long enough to replace itself via reproduction. Development founded on such objectives simply mimics ecological processes that shape the lives of all organisms, and its primary metric of success is population size. Human development, however, is not concerned with numbers or mass. Rather, it focuses on social objectives that are informed by the constituents of human well-being. The objectives of human development include longevity, health, education, and access to resources sufficient to allow one to make choices, pursue opportunities, and to meet or exceed socially established standards for a decent living. Within the context of food and farming, human development and human well-being require the production and consumption of a diverse array of foods to support a nutritious, balanced diet. This means that the energy content of food matches the activity levels of its consumers and that every essential nutrient is present in sufficient concentrations to ensure health. A nutritious, balanced diet, however, has to take into account freedom of choice, including freedom to choose what we produce and what we choose to consume. The ability to exercise such freedoms and choices requires ensuring access to what is needed to produce (seed, land, knowledge, and other forms of capital) and what is needed for a culturally appropriate, nutritious, and balanced diet. It is important to note, however, that human development and its objective of human well-being are agnostic on the issue of sustainability. Sustainable food and farming requires producing calories and nutrients sufficient to feed humanity such that we all lead a good life, one that is healthy and typified by freedoms and choices, but that also does not generate environmental harm that our children and others will inherit. Whether sustainable human development is feasible, from the standpoint of food and farming, is the topic of the next chapter.

5

BRINGING IT ALL TOGETHER

The Sustainable Future Before Us

THE ANTHROPOCENE AND AGRICULTURE

Humanity's Epoch

In chapter 3, we observed that humanity's impact on Earth is so massive that it is global in extent and rivals that of any other species that has ever lived on Earth. Because of this anthropic transformation of Earth, many have argued that we should name our current geological epoch the Anthropocene. Unlike other geological terms, such as "Pleistocene" or "Holocene," "Anthropocene" is not official yet, though the name is widely used. The International Union of Geological Sciences has the final say, and they are still deliberating, but it appears "Anthropocene" is on its way to becoming official.

Though the scientific process of establishing a new geological period seems an abstract exercise of interest mostly to geologists who concern themselves with naming eons, eras, periods, epochs, and stages (e.g., we live in the Phanerozoic eon, Cenozoic era, Quaternary period, Holocene epoch, and the Meghalayan stage), that the term is already commonly used points to the widespread sense that ours is a time that warrants special recognition.

Currently, one of the key debates concerns what, specifically, can be used to identify the Anthropocene in the paleorecord. If paleontologists of the future were to drill down into Earth's crust to the sediments and fossils being laid down today, what would tell them they have reached the Anthropocene? There are a number of proposed telltale signatures of the Anthropocene: mass extinction, a rapid rise in atmospheric CO_2, microplastic particles, a layer of airborne black carbon dust from fossil fuel burning, rapid rises in N and P attributed to industrialization, dominance of crop plant pollen, even the bones of broiler chickens.[1] In marine samples, it includes evidence of acidification and dramatic declines in coral, an anomalous abundance of plutonium-239 from nuclear fallout from aboveground nuclear weapons testing, and possibly other permanent signatures of humanity's impact.

Another key debate concerns when the Anthropocene actually started. Each candidate signature for the Anthropocene signals a different start date. For example, if we consider the disappearance of Earth's largest mammals as one of the first signs of anthropogenic mass extinction, then the Anthropocene started about 125,000 years ago. If we were to use the anomalously high acceleration of atmospheric CO_2 levels (~100 times higher than what has been observed in the last 500,000 years), then the Anthropocene would have started only about 11,000 years ago. Using a spike in plutonium-239 would date the Anthropocene to just 70 years ago. We have to decide when the Anthropocene began, and there are clearly a number of possible start dates. But perhaps that desire for a precise start date is unreasonable—an artifact of living within this time of dramatic global change.

The Onset of Agriculture

This primer suggests that the onset of agriculture is one of the better start dates for the Anthropocene because most of the tell-tale signs of humanity ultimately trace back to agriculture's origins. Although some argue for the more recent starting point of the Industrial Revolution (the late 1700s), when population growth, pollution, and mass extinction skyrocketed, the Industrial Revolution was ultimately fueled by agriculture. Agriculture allowed for different social organizations—including increased social stratification—and innovative development in scholarship, the arts, and research. By decoupling people from hunting, fishing, and gathering food from nature, agriculture triggered the development of urban concentrations whose citizens acquire food in markets. Being freed from the necessity of hunting and gathering food opened up space for various sociocultural developments, both positive, such as the establishment of institutes of learning, religion, and art, and negative, such as colonialism, slavery, and indentured servitude. It also allowed for the escalation of conflict to regional and even global wars. The impacts of the Industrial Revolution, also both positive and negative with respect to social development and the environment, are ultimately traceable to the onset and spread of agriculture as a means of meeting a population's food requirements.

The Best and Worst of Times

As we considered in chapter 1, the alterations we have made to the living world are nothing short of miraculous; they are testaments to our industry and ingenuity. Ecomodernists, for example, paint a largely positive picture of our niche reconstruction. While they note that nature has been devastated, they focus on the fact that humanity has flourished: longevity has doubled; the incidence of infectious disease has decreased, and successful treatments for once-lethal illnesses are abundant; we are more resilient to disturbances and extreme events such as fires, droughts, and famine; we experience less violence; democracy has largely displaced autocracy; and justice, freedom, equity, and tolerance are seen as on the rise. To quote from the "Ecomodernist Manifesto": "Historically large numbers of humans—both in percentage and in absolute terms—are free from insecurity, penury, and servitude."[2]

These positives have been unevenly distributed, however, and the environmental costs have been severe. Over 820 million people remain undernourished, 151 million children suffer stunting, and 51 million suffer muscle loss (wasting) from dietary deficiencies (chapter 4). Two billion people are suffering from micronutrient deficiencies, while 2.1 billion are overweight. Dietary diseases now comprise the largest disease burden humanity faces and pose greater risks of morbidity and mortality than unsafe sex, alcohol, and drug and tobacco use combined.[3] Agriculture has transformed between 40 and 60 percent of Earth's land surface, and it is the primary cause of contemporary mass extinction, is responsible for about a third of global greenhouse gas emissions, and accounts for almost three-quarters of all freshwater use. Marine systems may be more severely affected, with 60 percent of world fish stocks fully fished, and thus unable to be more productive, and a third overfished.[4]

Who is right about the state of the world, modernists or realists? Could we use a bookkeeper's approach to measure costs such as inequality, mass extinction, pollution, and climate change against the benefits of improvements in human longevity, food sufficiency, and the rise in democracy? It might seem like an objective approach, but there is no rational way to conduct such an analysis; thus, neither the modernist nor the realist view provides an accurate description of the state of the world if one considers the three pillars of sustainable development (chapter 3) and the conditions required for human development (chapter 4).

The Anthropocene is very much like the famous opening of Charles Dickens's novel *A Tale of Two Cities*, whose storyline is based on the events surrounding the massive upheaval of the French Revolution of the late 1700s:

> It was the best of times, it was the worst of times, it was the age of wisdom, it was the age of foolishness, it was the epoch of belief, it was the epoch of incredulity, it was the season of Light, it was the season of Darkness, it was the spring of hope, it was the winter of despair, we had everything before us, we had nothing before us, we were all going direct to Heaven, we were all going direct the other way.

In this concluding chapter, we suggest that the present is the best and worst time for humanity to pursue food sustainability. We build upon the foundations covered in chapters 1 through 4 and consider some of the key contemporary issues surrounding the sustainability of food, farming, and agriculture.

CHALLENGES AND BARRIERS TO SUSTAINABILITY

The Challenges

In each chapter thus far, we found many challenges to making food, farming, and agriculture sustainable. Here, we consider two cross-cutting challenges: the challenge of food inequity and the challenge of feeding a population headed toward 10 billion people by 2050.

THE CHALLENGE OF FOOD INEQUITY

In chapters 1 and 4, we used the World Health Organization's (WHO) recommended daily diet to provide a quantitative visualization of the challenges of producing the balance of calories and nutrients necessary for a healthy population of 7.6 billion people. Vast quantities of carbohydrates, vegetables, fruits, nuts, and water must be produced every day, yet their production is currently the source of considerable environmental harm. Many nations have outlined their own recommended diets, and one can ask, as Behrens et al. did,[5] whether the environmental consequences of people switching from their average daily diets to nationally recommended diets has had positive or negative environmental impacts. Though their study was limited to thirty-seven nations, their analyses suggest that the adoption of nationally recommended diets would reduce agriculture's negative environmental impacts, in terms of reduced greenhouse gas emissions, pollution, and land acreage used. This would seem like a "win-win-win," in the sense that people would be healthier, environmental impacts would lessen, and there would be a shift toward employing more sustainable agriculture practices. However, the biggest reductions in negative environmental impacts would be in richer nations, thanks to less consumption (i.e., kcals) and changes in diet composition, while poorer nations would suffer increased negative environmental impacts, primarily from the increased consumption of animal products (assuming that the increase in nutritive food was locally sourced).

This extreme inequity in food access and consumption is a major challenge to achieving sustainability within the agricultural sector. The production of a nutritionally balanced food supply to feed over 7.6 billion people is a staggering enterprise, yet it is one we are currently able to do, though the environmental costs are extremely high. The challenge is to reconfigure the social, economic, and environmental dimensions of this enterprise such that it ensures equitable access, distribution, and consumption and minimizes environmental costs. A tall order, to be sure, but at a fundamental level, this is what motivates transitioning to sustainable food, farming, and agriculture.

THE CHALLENGE OF FEEDING 10 BILLION BY 2050

If we imagine that we can overcome the current impediments to establishing a sustainable food system that eliminates undernourishment and nutritional disease worldwide, as seems within our grasp, we then have to move to the next question: Is such a system truly sustainable in the face of population growth? More simply, can we feed 10 billion by 2050?

Foley and colleagues attempted to answer this question in 2011.[6] They asked what it would take to solve the problems of dietary inequality around the world without worsening environmental conditions—the sine qua non of sustainable agriculture. Their solutions were to (1) stop expanding agriculture to retain nature for the critical services it provides (chapter 2), (2) improve yields where farms are underperforming (i.e., close the YIELD GAPS), (3) increase resource use efficiency (i.e., use only the fertilizers, biocides, and water necessary to obtain maximum sustainable yields), (4) shift diets (e.g., less omnivory and more vegetarianism), and (5) reduce waste (currently 40 percent of the food harvested in developing countries goes to waste; in industrialized countries, 40 percent of food is wasted at the consumer level). Their analysis and objectives are shown in figure 5.1.

Surprisingly, especially to many who felt there is no way increases in agricultural production can keep up with exponential human population growth, Foley et al. said the answer to

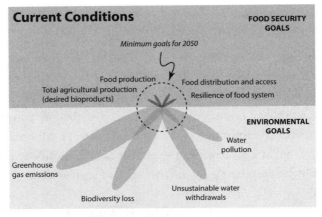

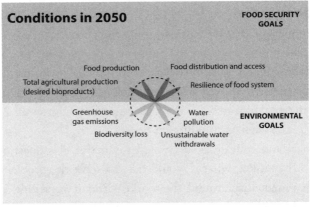

FIGURE 5.1 Feeding the world's population in 2050.

Foley and colleagues addressed the question of whether we can feed the world population in 2050, which is estimated to be ~10 billion. They illustrated the question using flower-petal diagrams, which are useful ways to visualize multiple factors in a single graph. In the top figure, for four key food security goals and four critical environmental goals, the authors concluded that currently we are meeting food security goals (i.e., the tips of the petals are below the dashed circle describing the minimal acceptable conditions), but environmental goals are not being met (i.e., petal tips are well above the dashed circle). Ideally, all the petal tips would be within the dashed circle by 2050, in spite of the significant increase in population. Surprisingly, the authors concluded that these food security and environmental goals can be met by 2050 provided a number of steps are taken, such as closing yield gaps, reducing waste, and having people shift their diets to consuming less meat.

Source: J. A. Foley, N. Ramankutty, K. A. Brauman, et al., "Solutions for a Cultivated Planet," *Nature* 478 (2011): 337–42.

the question of whether we can feed 10 billion by 2050 is *yes*. The conclusions of Foley et al.'s analysis, however, ignored the social dimensions of the solutions they proposed. The authors were upfront about this, saying, "Our analysis focuses on the agronomic and environmental aspects of these challenges, and leaves a richer discussion of associated social, economic and cultural issues to future work."

Foley et al.'s analysis was prompted by the challenge of feeding 10 billion people by 2050 and used existing data and models to answer the question. Their analysis is only one of many analyses in an increasingly popular and crowded field of researchers and practitioners trying to answer this question. The basic approach is to establish environmental constraints, usually allowing for no further deforestation, and then explore what happens when one varies agricultural factors such as yield gaps, waste, the use of genetically modified crops (GMOs), industrializing agriculture, biocide use, irrigation, and more. One of the more interesting exercises in this feed-10-billion-by-2050 endeavor is that of Erb et al.,[7] in which the authors used a computer model to vary agricultural factors in 500 different deforestation-free scenarios. They found that two-thirds of these 500 individual scenarios supported the possibility of feeding 10 billion by 2050, which suggests that there are many ways to achieve this goal without expanding agricultural lands at the expense of forests and the services they provide. Thus, Foley et al.'s solutions may not be as unrealistic as they first seem.

The answer to the question of whether we can feed 10 billion by 2050 appears to be *yes*, but with the major caveat that there can't be any social barriers to adopting and implementing the solution(s) we devise. Given the absence of analyses that include social dimensions, it thus remains unclear if the answer is actually *yes* or *no*.

The Barriers

Tackling the major challenges of sustainability, in particular achieving food equity across the world and feeding 10 billion in 2050, is impeded by a number of barriers. Here we consider three of the more significant ones: industrial dependency, divergent narratives, and conflicting policy frameworks.

SOCIAL BARRIERS TO SUSTAINABLE FOOD, FARMING, AND AGRICULTURE

Unlike many analyses of the challenge of feeding 10 billion by 2050, the International Panel of Experts on Sustainable Food Systems (IPES-Food)[8] integrated across social and natural dimensions to consider what factors in food, farming, and agriculture enable or impede sustainable development. Based on seven case studies, they found that social change, such as community-led governance; forging partnerships among researchers, farmers, and environmental groups; and empowering women and youth can move food and farming systems toward greater sustainability. However, social factors, especially those related to economic and political bias toward industrial agriculture, impede the spread of sustainable practices.

IPES-Food identified several barriers to transitioning to sustainable food and agriculture. These are:

- Increased efficiency of scale decreases agrobiodiversity
- Emphasis on production favors unsustainable industrial agriculture
- Agricultural development focused on export to global markets favors developed nations over less developed ones
- Societal expectations by industrialized countries that food should be low cost, despite its high social and environmental

costs, forces developing countries to provide low-cost food and absorb the environmental costs

- High rates of agricultural production are favored by the short time horizons of politics (e.g., term limits), research (e.g., 3-to-5-year funding cycles), and business (e.g., expectations of returns on investments to be manifested over one to a few years)

- Compartmentalized approaches to development rather than integrated approaches (e.g., Sustainable Development Goals [SDGs] that compartmentalize objectives regarding water, food, terrestrial systems, and marine systems into separate goals, as will be discussed in what follows) makes developing holistic solutions challenging

- The use of metrics of success that focus on wealth rather than sustainability (e.g., monetary metrics for poverty rather than measures of human well-being or policies based on calories rather than nutrition)

Such social barriers are not insurmountable, however, and many organizations promote sustainable food, farming, and agriculture as a means to overcome these barriers. That said, progress is slow.

Divergent Narratives

One of the barriers to transitioning to sustainable agriculture is the many narratives associated with food, farming, and agriculture. Narratives are how we link cause and effect in order to tell a story or convey an idea. The multiplicity of narratives in agriculture produces divergent solutions. Multiple divergent narratives in food sustainability can be a barrier to pursuing sustainability if they promote different policies, actions, or prioritizations in planning.

Béné et al. illustrate the problem of divergent narratives by comparing different failure narratives associated with food systems.[9] The starting point is the premise that agriculture in its current form has failed us. This failure narrative has multiple divergent alternative narratives associated with it. Agriculture's failure can be viewed as being unable to feed future generations (i.e., no, we cannot we feed 10 billion by 2050), its failure to provide healthy diets, its failure to provide equal and equitable distribution of food, or its tendency to exact enormous environmental costs (figure 5.2). The overarching narrative concerns the failure of conventional agriculture, but each failure storyline highlights different urgencies that need to be addressed (e.g., closing food gaps, improving diet quality, supporting local autonomy, or protecting the environment, as outlined in figure 5.1).

Underdeveloped Policy Frameworks

Given the challenges and barriers to developing a sustainable food system, one needs a framework that can guide individuals, communities, nations, and global organizations to devise policies that enable sustainable food, farming, and agriculture. Two policy frameworks that reflect the outcome of consensus-driven exercises are the United Nations 2030 Agenda for Sustainable Development and its associated goals and "Food in the Anthropocene,"[11] led by the EAT-Lancet Commission. While both frameworks address sustainable development and human well-being and represent steps forward on the path to facing global challenges, they each have their limitations.

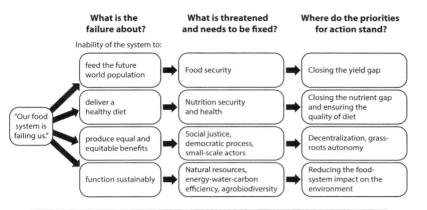

| What is the failure about? | What is threatened and needs to be fixed? | Where do the priorities for action stand? |

Inability of the system to:

"Our food system is failing us."

feed the future world population → Food security → Closing the yield gap

deliver a healthy diet → Nutrition security and health → Closing the nutrient gap and ensuring the quality of diet

produce equal and equitable benefits → Social justice, democratic process, small-scale actors → Decentralization, grass-roots autonomy

function sustainably → Natural resources, energy-water-carbon efficiency, agrobiodiversity → Reducing the food-system impact on the environment

FIGURE 5.2 Divergent narratives associated with food systems and why they fail.

Transitioning to environmentally sustainable food systems is hindered by the confusion multiple narratives can create. This figure provides an example of four divergent *failure narratives*, or narratives about why food systems fail. Each narrative defines failure differently, each identifies different aspects of the food system that are failing, and each promotes different actions to redress or prevent food system failure. Because the narratives are divergent, they can generate confusion and lead to different outcomes depending on which narrative stakeholders and decision makers subscribe to. What is needed are convergent narratives that lead to a coherent understanding of sustainable food systems.

Source: Authors.

The UN 2030 Agenda for Sustainable Development

The United Nations, following on from the Millennium Development Goals, officially launched the Sustainable Development Goals (SDGs) in 2016 (after adopting the 2030 Agenda in 2015). There are 17 goals (box 5.1), 169 targets, and 232 indicators, with 2030 as the date by which all of the goals should be achieved. In the SDG framework, each goal broadly describes a critical outcome that must be achieved if we are to transition successfully from unsustainable to sustainable development. Each SDG is divided into more narrowly defined, specific targets, and progress toward each target is assessed using indicators, which are explicit, measurable variables directly related to each target. For example, goal 2, entitled "End Hunger," consists of eight targets. A few of these targets are (1) to ensure access to safe, nutritious, and sufficient food; (2) end all forms of malnutrition; and (3) double agricultural productivity and the incomes of small-scale food producers. For the first target, indicators include (1) quantification of the prevalence of undernourishment in a population and (2) quantification of the prevalence of food insecurity using the Food Insecurity Experience Scale, an index developed by the Food and Agriculture Organization of the United Nations (FAO). We refer the reader to the UN website for a complete list of goals, targets, and indicators,[12] but the example of SDG2 provided here illustrates the ambitious nature of the goals as well as the seemingly arbitrary nature of the targets and indicators. For example, why indicator X rather than indicator Y? Though the SDGs may seem like a mixed bag of policy guidelines, they were the end product of three years of deliberations by all the world's nations and are an expression of remarkable consensus on development goals.

The first thing to note about the SDGs presented in box 5.1 is their obvious connections. For example, SDGs 1, 2, and 3 (poverty, hunger, and health, respectively) are all tied to one another. The poor often cannot afford safe, healthy, nutritious, and sufficient quantities of food; thus, malnutrition and hunger are common among the poor, and poor health is an inevitable consequence of the poor diets of the impoverished. A policy framework must see these (i.e., SDGs 1, 2, and 3) as

BOX 5.1 THE UN SUSTAINABLE DEVELOPMENT GOALS (SDGS) AND FOOD, FARMING, AND AGRICULTURE

The UN established seventeen SDGs to be met by 2030:

1. No Poverty: End poverty in all its forms everywhere
2. Zero Hunger: End hunger, achieve food security and improved nutrition, and promote sustainable agriculture
3. Good Health and Well-Being: Ensure healthy lives and promote well-being for all at all ages
4. Quality Education: Ensure inclusive and equitable quality education and promote lifelong learning opportunities for all
5. Gender Equality: Achieve gender equality and empower all women and girls
6. Clean Water and Sanitation: Ensure availability and sustainable management of water and sanitation for all
7. Affordable and Clean Energy: Ensure access to affordable, reliable, sustainable, and modern energy for all
8. Decent Work and Economic Growth: Promote sustained, inclusive, and sustainable economic growth; full and productive employment; and decent work for all
9. Industry, Innovation, and Infrastructure: Build resilient infrastructure, promote inclusive and sustainable industrialization, and foster innovation
10. Reduced Inequalities: Reduce inequality within and among countries
11. Sustainable Cities and Communities: Make cities and human settlements inclusive, safe, resilient, and sustainable
12. Responsible Consumption and Production: Ensure sustainable consumption and production patterns
13. Climate Action: Take urgent action to combat climate change and its impacts

14. Life Below Water: Conserve and sustainably use the oceans, seas, and marine resources for sustainable development
15. Life on Land: Protect, restore, and promote sustainable use of terrestrial ecosystems; sustainably manage forests; combat desertification; halt and reverse land degradation; and halt biodiversity loss
16. Peace, Justice, and Strong Institutions: Promote peaceful and inclusive societies for sustainable development; provide access to justice for all; and build effective, accountable, and inclusive institutions at all levels
17. Partnerships for the Goals: Strengthen the means of implementation and revitalize the global partnership for sustainable development

Only goal 2 is directly related to sustainable agriculture, but several goals are indirectly related. Poverty determines one's ability to secure food (goal 1); nutrition is critical to health (goal 3); women play key roles in food and agriculture, both as owners of farms and as laborers, and thus a sustainable food system is critical for achieving women's economic equality (goal 5); 70 percent of freshwater withdrawals are for agriculture (goal 6); agricultural development is linked with economic development (goal 8); reducing food waste is essential to achieving sustainability within the contexts of production and consumption (goal 12); agricultural practices and land use contribute to climate change, and climate change impacts the agricultural sector (goal 13); and food, farming, and agriculture are the primary contributors to environmental degradation on land and in the sea (goals 14 and 15).

inextricably linked, not as separate objectives. The same can be said for all seventeen SDGs.

Le Blanc conducted a formal network analysis of the SDGs by identifying links through shared targets.[13] The result is a picture of the SDGs as a web of linked goals (figure 5.2), in contrast to the linear presentation described in box 5.1. Figure 5.3 reveals serious deficiencies in the SDGs as a sustainable development policy framework. For example, SDGs 14 and 15, marine and terrestrial ecosystems, respectively, are not linked

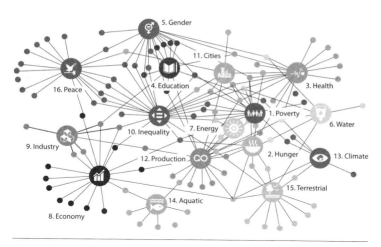

FIGURE 5.3. The Sustainable Development Goals (SDGs) illustrated as a network of linked objectives.

Using an analysis by Le Blanc (2015), the SDGs are shown to be a network in which each SDG represents a node linked to the others by their shared objectives. What is important to note is that the SDGs are linked to one another in multiple ways, meaning that they cannot be achieved in isolation. The entire group of SDGs has to be pursued as an integrated set.

Source: David Le Blanc, "Towards Integration at Last? The Sustainable Development Goals as a Network of Targets," *Sustainable Development* 23, no. 3 (2015): 176–87.

to one another in spite of the physical, chemical, biological, and innumerable environmental connections between land and sea. Pollution from agricultural runoff at the mouths of major rivers, for example, leads to dead zones in the sea. Hunger (SDG2) in figure 5.2, on the other hand, is correctly linked to poverty (SDG1), health (SDG3), water (SDG6), gender (SDG5), land (SDG9), employment (SDG8), and sustainable consumption and production (SDG12) but not to marine systems (SDG14), climate change (SDG13), or sustainable industrialization (SDG15). Marine systems (SDG14) are also not linked to climate change (SDG13), in spite of the enormous role oceans play in climate regulation and the massive impacts climate change has on marine systems.

In fact, as this primer has emphasized throughout, since all aspects of sustainable development and human well-being ultimately trace back to food, farming, and agriculture, the SDGs as a policy framework should reflect the centrality of food sustainability to all its goals. Murray, in collaboration with the Economist Intelligence Unit, provided an alternative structure for the SDG network.[14] In this network, "food is a common thread linking all 17 UN Sustainable Development Goals (SDGs), given the interconnected economic, social and environmental dimensions of food systems." Sustainable food systems sit at the center of a Venn diagram in which SDGs are grouped into the three pillars of sustainable development (figure 5.4).

While it is laudable that there are global commitments to well-defined sustainable development goals, targets, indicators, and a specific timeline, social-economic-environmental integration remains underdeveloped. The good news is that much progress is being made to provide integrative guidelines to implement the SDGs where food, farming, and agriculture play a pivotal role, as seen in figure 5.3. Further, the

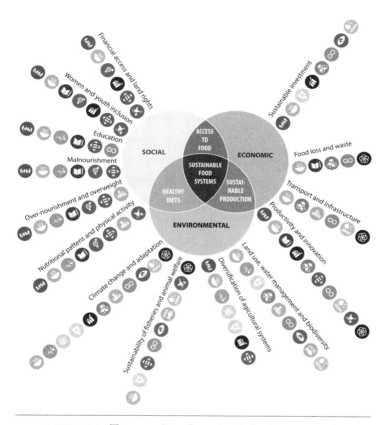

FIGURE 5.4. The centrality of sustainable food systems to the Sustainable Development Goals grouped by their social, economic, and environmental targets. Compare this approach to the network shown in figure 5.2.

Source: After Sarah Murray, "Fixing Food 2018: Best Practices Towards the Sustainable Development Goals," ed. Martin Koehring, Economist Intelligence Unit, 2018, https://foodsustainability.eiu.com/whitepaper-2018/.

UN 2030 Agenda for Sustainable Development points to the critical necessity to integrate across the SDGs, and tools are available for all sectors to develop and implement integrative policies.[16]

It is important to note that establishing goals for sustainability is just a first step. For example, to achieve a sustainable food system requires

- implementable policy (e.g., developing policy consistent with the economics, capacity, and law of the region),
- socially congruent behavior (e.g., there is widespread subscription to the ideals underpinning food sovereignty, justice, and security), and
- a political economy (e.g., a political process that influences production and economic outcomes) that promotes sustainability.

Decision making by farmers, which is where farming practices in a region are determined, is strongly influenced by the social, cultural, and policy environments farmers find themselves in. If these environments do not promote sustainability, they can create barriers to achieving sustainable development goals.

Food in the Anthropocene

In contrast to the SDGs, the EAT-Lancet Commission issued a report in 2019 that echoed the analysis by Foley and colleagues described earlier; that is, it is another version of the feed-10 billion-by-2050 narrative. The commission claimed to be a neutral, nonpartisan, independent collection of thirty-seven experts in human health, agriculture, political science, and environmental sustainability.[17] The commission begins with the argument that "food in the Anthropocene represents one of the greatest health and environmental challenges of the 21st century," thus placing food at the center of the framework. It proposed a new global agenda it calls the Great Food Transformation.

The EAT-Lancet report is one of win-win optimism, claiming that healthy diets can be provided for everyone by 2050. There are some serious caveats, however. Such a transformation requires that humanity undergo a dramatic global change in its approach to food, farming, and agriculture. Of the many changes that must occur is a global dietary shift that requires a staggering reduction in the consumption of foods such as red meat and sugar and a doubling of the consumption of fruits, nuts, and vegetables. It will also require transforming agriculture practices to reduce yield gaps by 75 percent. Such a global transformation necessitates participation by just about every person, community, nation, and institution in the world. Steps include securing global commitments by all nations to shift their diets to vegetarianism, shift their current emphasis on yields to emphasize healthy food, abandon industrial intensification in favor of sustainable intensification, increase global cooperation on the management of land and sea, and cut food loss and food waste to half of what it is today. Further, such commitments must result in actions that keep humanity within the planetary boundaries framework (chapter 3). The commission acknowledged that the Great Food Transformation, as it is called, is uncharted policy territory and faces obstacles that are not easily overcome.

RETHINKING, REINVENTING, AND REINVIGORATING FOOD PRODUCTION AND FARMING

Given the best and worst of times we find ourselves in, it is important to keep sight of advances arising from the best of times (e.g., social and scientific innovations) that promise to counter the worst of times (the dire state of the environment). The first questions to ask, before considering how food, farming, and agriculture are to be reinvented, are: What kind of world do we want to live in? Do we want to take a business-as-usual approach and risk crashing and burning, or do we have an end goal, a known world that we want, that can guide what we need to change now to arrive at a better future? In light of where we want to go, what are the current aspects of food, farming, and agriculture that will facilitate a sustainable future? There are many topics in this area, but technological advances, organic farming, sustainable intensification, and the recognition of the importance of both small- and large-scale farming are central to today's views on sustainable food systems.

The World We Want

If good global governance (i.e., enabling policy frameworks) is what is necessary for achieving environmental sustainability, what should guide its objectives, given the extraordinary diversity of peoples around the world? In 2013, in order to better inform the Sustainable Development Goals, the United Nations conducted a global survey of 194 nations and over 1 million people to find out what sort of world people want.[18] Not surprisingly, better education, health, water, sanitation, gender equity, and ending poverty and hunger were at the top of the list. The respondents also wanted an end to the pervasive growth in inequality, where more and more people receive less and less benefit from development while a tiny fraction profits enormously. The list of the victims of inequality was long, including the rural poor, urban slum dwellers, women and girls, people living with disabilities, indigenous people, migrants, displaced people, and those marginalized because of religion, ethnicity, or sexual orientation. In sum, the survey found that the world the majority wanted is one where "governments . . . do a better job . . . delivering key services, encouraging growth while regulating markets, and preventing insecurities associated with compromising the planet and the wellbeing of future generations."

The conclusion drawn from the survey was that the world we want is one that is environmentally sustainable and socially just. As we have argued throughout this primer, this world can only be achieved if we can make food, farming, and agriculture environmentally sustainable.

The Rise of Organic Farming

Organic farming is an old idea, one traceable to Sir Albert Howard, F. H. King, and Rudolf Steiner in the 1900s and based on traditional techniques from around the world. Organic farming is, broadly speaking, farming without the use of inputs typically associated with conventional agriculture and derived from synthetic sources, such as inorganic fertilizers from industrial or mining operations, synthetic biocides, and, more recently, genetically engineered organisms. Organic farming suffers from many misconceptions by both its advocates and its detractors. Advocates often uncritically equate organic farming with sustainability and see it as a panacea for the many ills industrial farming has wrought upon the planet. Detractors often uncritically equate organic farming with the low-yield production of high-priced goods the poor cannot afford and, noting its larger footprint than conventional farming, arguing that it is a threat to remaining natural habitats. In spite of its detractors, organic farming is on the rise, with sales of organic products more than quintupling since 1999 and sustainability policies and strategies increasingly becoming prevalent in public discourse.

Detractors are right that organic farms are indeed less productive than conventional farms. In support of the advocates, however, organic farm yield varies enormously among its products. Yield gaps, or the gap between organic and conventional farming, are often around 19 percent—organic farms, on average, produce about a fifth less than conventional farms. In some cases, however, there is no significant yield gap (e.g., for some leguminous and perennial crops, especially where diversified farming is practiced). Yield gaps for other crops can be as little as 6 to 11 percent (e.g., rice, maize, soy, and clover) or as high as 28 percent (e.g., for fruits, vegetables, and wheat). Thus it is not true that all organic farming is always prohibitively lower in productivity. Organic products also do indeed command higher prices, but this occurs primarily where price premiums are in place (e.g., higher prices for organically certified products). Yield gaps and pricing aside, the environmental and health benefits of organic production are well documented.

Comparing conventional with organic farming is clearly complicated, and there are legitimate arguments for and against different aspects of the practice. In their review of conventional versus organic farming, Reganold and Wachter (2016) argue that, from a holistic perspective in which one includes social, economic, and environmental costs, organic farming exceeds conventional farming as a sustainability solution on many fronts (figure 5.5).[19] Another observation is that conventional farming has benefited enormously from investments and subsidies by governments, agribusinesses, and agronomic institutions. If similar investments were made to support organic agriculture (e.g., for research), it may be possible to reduce yield gaps and organic product prices.

There have been several prominent reviews of conventional versus organic farming, with complex findings,[20] and the debate continues. As is often the case with dichotomies in worldviews,

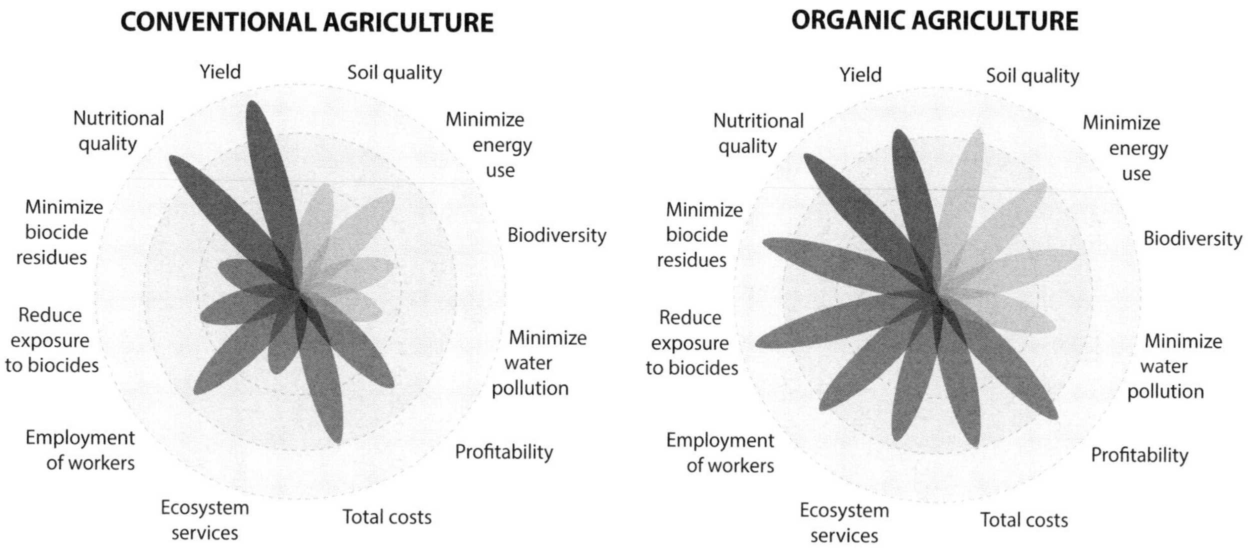

CONVENTIONAL AGRICULTURE

ORGANIC AGRICULTURE

FIGURE 5.5 Organic versus conventional farming.

Using the same flower-type graphical approach taken in figure 5.1, this figure compares conventional and organic farming across twelve key factors. In this figure, the longer the petal, the better its performance according to sustainability criteria. Based on qualitative assessments of a large number of studies, the figure shows organic farming to be the more sustainable option (i.e., most of its petals are similar in length and, on average, longer than those of conventional farming).

Source: After John P. Reganold and Jonathan M. Wachter, "Organic Agriculture in the Twenty-First Century,"
Nature Plants 2, no. 2 (2016): 15221.

pitting organic against conventional farming is counterproductive. Rather than a vigorous debate, rethinking agriculture is likely to find that the best sustainability solution is a mixed strategy, one where farming uses best practices drawn from both conventional and organic farming.

Sustainable Intensification

SUSTAINABLE INTENSIFICATION at first sounds like an oxymoron (intensification seems the opposite of sustaining the norm), but there has been considerable interest in forms of agricultural intensification that can lower yield gaps without incurring the environmental costs common to conventional agriculture. Over recent years, several national and international organizations have adopted sustainable intensification in their policies and planning for achieving food sustainability, but without appraisal of the plausibility, feasibility, scientific basis, or evaluation of alternative approaches.

What constitutes sustainable intensification is a contentious subject,[21] but if one takes the broad perspective that it involves producing more food than an acre typically produces without significantly harming the environment, many practices fall under this banner, including

- land using integrated pest management, which simultaneously optimizes pest control while minimizing environmental harm and may involve a mix of biocides, biocontrol, use of GMOs, or other pest management strategies;
- conservation agriculture, which maximizes production while minimizing soil degradation and maximizing biodiversity conservation;

- diversified farming;
- well-managed, biodiverse forage and pasture lands with live-stock rotation;
- agroforestry, including the cultivation and conservation of trees in agricultural systems;
- smart irrigation, including well-managed watersheds and the use of irrigation technology (e.g., drip irrigation) to minimize runoff; and
- landscapes that include intensive small- and patch-scale agricultural systems such as community farms, vertical farming, and backyard gardens.

Under such a broad definition, Pretty and colleagues estimate that there are an estimated 163 million farms (about a third of all farms) in over a hundred countries engaged in sustainable intensification, equivalent to an area of nearly a billion acres (450 million ha) of agricultural land (about 9 percent of the total area under cultivation).[22]

Critics of sustainable intensification, however, feel that such approaches repeat the errors of the Green Revolution by focusing on production. In contrast, sustainability seeks to match production with demand within the constraints of environmental sustainability. Matching food production with demand involves two sides—the demand side and the supply side. As we have seen in the feed-10-billion-by-2050 narratives and the Great Food Transformation, both call for dietary shifts (a demand-side factor) and closing yield gaps (a supply-side factor). Uncritical adoption of sustainable intensification focuses on the supply side, when multiple analyses have shown that sustainability requires changes in both supply and demand. Furthermore, the idea of sustainable intensification disregards the economic and social dimensions of sustainable development. Some argue

that the greater urgency for sustainability is not in intensification or greater production per unit area; rather, the urgency is
greater for food security, food justice, food sovereignty, and for
meeting the many social objectives identified in the World We
Want survey and the SDGs.

The Optimism of Innovation and Technology

A big part of sustainable intensification is founded on research,
innovation, and advances in technology that promote maximal
food production while minimizing or completely avoiding harmful environmental impacts. There is a sense that innovation and
the creation of new technologies will always emerge in the nick
of time and rescue us when crises arise (box 5.2). Historically,
the development of the Haber-Bosch process for making nitrogen fertilizer in the early 1900s and the multiple scientific
advances that led to the Green Revolution seem to support this
view. Innovation and technology in agriculture have, indeed,
allowed production to keep pace with population growth in spite
of constant concerns that it never could. In reality, however,
innovations in development are not predictable.[23] The emergence
of innovations involves many people and many institutions across
many scales, from the individual to the global, and can face many
barriers to development, dissemination, and adoption. When it
comes to sustainable development, innovation is further ham-
pered by current financial, regulatory, and reward systems that
rarely serve the poor, let alone the more abstract future generations whom sustainable development is supposed to serve. Many
innovations have also proven to have tradeoffs, as is abundantly
evidenced in many of the advances associated with the Green
Revolution that had enormous, unintended, and unexpected
environmental and social costs.

BOX 5.2 TECHNOLOGY AND
SUSTAINABLE AGRICULTURE

A number of modern technologies in farming and agriculture offer to reduce fertilizer, water, energy, and biocide usage while improving yields and food quality without expanding agriculture's footprint; thus they are often associated with sustainability. While there is much promise associated with innovation in agricultural technology, these products and techniques are often too expensive for smallholder farmers, regulated in ways that favor developed over less developed countries, cannot be adopted by countries that lack the capacity to employ new technologies, and can be associated with novel hazards, such as the environmental hazards associated with the toxicity of nanomaterials or the unwanted spread of genes in genetically modified organisms (GMOs) to nontarget organisms. Here we list just a few emerging technologies associated with sustainability.

Precision agriculture: Management that uses information technologies to bring data from multiple sources to bear on decisions associated with agricultural production. This technology includes equipment and tools such as self-driving tractors, geographical positioning systems (GPS), remotely sensed images (from drones or satellites), soil-water sensors that can be used to regulate irrigation, and wearable fitness trackers placed on cows to locate and monitor their activity and health.

Nanotechnology: Technology involving materials that are about 1 to 100 nanometers in size. With respect to food, they may be used to improve taste, flavor, color, and texture; to create packaging that has greater structural integrity and antimicrobial properties; or to microencapsulate nutritional supplements. Nanomaterials are also used to engineer microsensors for monitoring environmental conditions. Nanoscale materials include fertilizers, biocides, and growth hormones that allow for greater precision in administration.

'Omics: Methods that involve the direct manipulation of the molecular biology of organisms are collectively referred to as 'omics because of their relation to the study of an organism's DNA (genomics), its proteins (proteomics), its RNA derived from transcription of its genes (transcriptomics), the molecular biology of its metabolism (metabonomics), the molecular biology of its carbohydrates (glycomics), the molecular biology of its lipids (lipomics), and the simultaneous study of multiple genomes (metagenomics). 'Omics are data-intensive sciences that can be used to discover or search for genes in organisms that can provide the bases for genetic engineering of crops, livestock, and even microbial species in agricultural systems. Genomes for major agricultural species include rice, wheat, maize, potatoes, tobacco, tomato, manioc, the poplar tree, chicken, pig, cow, duck, and a number of agricultural microbes, including yeast, *Rhizobium* (nitrogen-fixing bacteria), pathogenic fungi, bacteria, and a number of viruses. These methods provide new ways to more precisely and knowledgably alter the genetics of agricultural species in ways that complement traditional breeding.

Gene editing: The use of molecular tools to make directed, targeted changes to an organism's genes, usually involving CRISPR-Cas9 (which stands for clustered regularly interspaced short palindromic repeats and CRISPR-associated protein 9). CRISPRs are short stretches of DNA (discovered in bacteria), and Cas9 is an enzyme capable of cutting DNA strands. The bacteria use CRISPRs as a form of viral immunity, in which CRISPR-Cas9 identifies and severs DNA sequences specific to invading viruses. These tools allow for precise genetic engineering of organisms, such as crops and livestock.

Counting on the timely emergence of innovation and technology to save us from crises in food, farming, and agriculture is a misguided and unreliable path forward for sustainable development.

Integrating Small and Big Farms

One of the clear tensions in food, farming, and agriculture is that of spatial scale, where the vast majority of the world's 570 million farms (84 percent) are smaller than 5 acres (1–2 ha) and constitute only 12 percent of the world's farmland. On the other hand, only 1 percent of the world's farms are large, but they collectively occupy 65 percent of the world's farmland (chapter 3). The tension occurs because large holdings tend to be industrial farms that are favored by world markets, agribusiness, and governments; this disadvantages the poor and gives rise to countermovements that promote food sovereignty, food justice, and food security to address such large-farm biases. In support of large farms, however, some argue that they produce cheap food desperately needed by the urban poor. In rebuttal, others will point to the poor quality of industrially produced cheap food, which adversely affects health both through direct consumption of the food and as a consequence of industrial practices that negatively impact health (e.g., water, soil, and air pollution).

Herrero et al. took a closer look at the relative contributions of different-sized farms to food and nutrition.[24] The authors found that small and medium farms (under 125 acres or 50 ha) produce 51 to 77 percent of nearly all the farm products and nutrients they examined. This finding varied considerably around the globe. In the Americas, Australia, and New Zealand, large farms dominated. In Europe, West Asia, North

Africa, and Central America, medium-size farms (50–124 acres, or 20–50 ha) dominated. In sub-Saharan Africa and Asia, small farms (under 50 acres or 20 ha) dominated, and in sub-Saharan Africa, Southeast Asia, and South Asia, very small farms (under 5 acres, or 2 ha) contribute about a third of the food and nutrients consumed by people globally. At the landscape level, biodiverse landscapes were also the dominant source of the majority of vegetables (81 percent), roots and tubers (72 percent), pulses (67 percent), fruits (66 percent), fish and livestock products (60 percent), and cereals (56 percent), as well as the majority of micronutrients (53 to 81 percent %) and protein (57 percent). Low-diversity landscapes (where large farms are often found), however, provide the majority of sugar (73 percent) and oil crops (57 percent) and thereby account for the majority of global calorie production (56 percent). A shift to a more balanced diet, one emphasizing vegetables, fruits, and nuts (i.e., a plant-based diet), will naturally draw upon the expertise and production of the farms producing these dietary components (i.e., smaller-scale, diverse farms) and thereby reduce society's reliance on the large farms producing food that is less nutritious and less supportive of human development and well-being.

CONCLUSION

We began with the notion that humanity is the single most successful species Earth has ever known, at least in terms of biological metrics like the number of individuals, the total mass of our species, the extent of our dominance of ecosystems across the globe, and the species-specific commandeering of natural resources. So extraordinary is our success and impact that the proposition to name this epoch in Earth's history the Anthropocene has found wide appeal.

Once we explored how nature works, however, we discovered that the costs of our success have taken their toll, producing a world that cannot continue long into the future unless we change how we do things. Global change, including climate change, emerging diseases, the spread of invasive species, pollution, and mass extinction, all tell us that things are dangerously unstable. All of Earth's ecosystems are undergoing massive change, and there is a real possibility of crossing tipping points and entering new, alternative stable states far less conducive to human well-being.

An exploration of the basics of how nature works, however, can help illuminate the issues, both the problems and the solutions. How elements combine to produce biomolecules; how biomolecules and other compounds produce the macro- and micronutrients essential to healthy human diets; and how 8.7 million species of plants, animals, and microbes play vital roles in producing stably functioning ecosystems, including agroecosystems, and the resilient biosphere that is our home, constitute the environmental basics of ecosystem functions. Traditional economic development, especially conventional farming over the previous two centuries, has altered the natural world to enhance the production of food, but in so doing, it has diminished other functions and services nature provides. What is needed is to transition from the traditional development of the last two hundred years to sustainable development, an idea that took hold in the 1980s and 1990s and has become the dominant model for development in this century.

Our focus in this primer has been on agricultural development, but we have seen that it is inextricably linked to all aspects of sustainable development. This is clearly seen in the

Sustainable Development Goals, where sustainable food systems are central to achieving all seventeen SDGs.

Farms, agriculture, and food systems, however, are complex milieus of natural, social, and economic processes. While natural systems have checks and balances that ensure resilience, robust and equitable human social systems rely on the adoption of the principles of food justice, security, and sovereignty as well as on cultural and political economic processes that promote sustainability. The nearly universal commitment by the world's nations to the SDGs is an important first step, but when it comes to achieving these goals, the way forward requires overcoming several challenges. It is encouraging that analyses suggest that we can feed 10 billion people by 2050 without further environmental harm, but it is not at all clear that crowded urban enclaves dominated by impoverished vegetarians existing at subsistence levels is the world we want. It is also not clear that the world currently envisioned for 2050 is sustainable to 2100 or beyond. The "world we want" as assessed by the United Nations is a different world, and the goals we have set for ourselves in the SDGs clearly reflect that.

While the SDGs describe a world that seems untenable given our current knowledge of how the world works, there is much room for optimism. New tools, new technology, new approaches in economics, new farming methods and practices, dramatic increases in knowledge concerning food sustainability, advances in environmental biology, widespread information exchange, and much more point the way to a desirable sustainable future if we have the will and motivation to go there.

We hope this primer provides a guide to what is ultimately an enormously complex issue but arguably the single most important one that we must address: to make food, farming, and agriculture environmentally sustainable.

SUMMARY: THE SUSTAINABLE FUTURE BEFORE US

In this, the Anthropocene, we find ourselves awakened to the realization that the biosphere, our birthplace and home, has changed dramatically from one that was diverse, resilient, and robust to one that is degraded, impoverished, and unstable, putting our record of steadily improving human well-being at risk. Unsustainable food, farming, and agriculture are the sources of our prosperity and the sources of the harmful environmental changes that characterize the Anthropocene. Today's world is not the world we want. There is a global desire for improvements in the social, economic, and environmental dimensions of well-being and for people to be able to live lives that they value. That is, there is a global desire to further sustainable human development. The way forward, given the topic of this primer, is to shift from unsustainable agricultural practices to sustainable ones that minimize environmental harm and are socially acceptable and economically viable. Technology; organic farming; sustainable intensification; an increasing appreciation of both small and large farms' contributions to local, regional, and global food systems; and much more are catalyzing transformations to sustainable food systems. There are also numerous local and global commitments to sustainability and sustainable development. However, the lack of a strong focus on sustainable food systems, the presence of too many divergent narratives, and a lack of practical integration across the social, environmental, and economic dimensions of development impede our progress toward a sustainable future. These commitments could be stronger, more cohesive, and more integrative, but on the whole, in spite of their shortcomings, they are moving us in the right direction—toward a food system that will eventually be environmentally sustainable, equitable, and aligned with pursuit of human well-being.

GLOSSARY

AGRICULTURE: See box 1.1.

AGROBIODIVERSITY: The array of species on and off farms or planned (e.g., crops and livestock) and unplanned (e.g., native species, exotics, or other species).

AGROECOLOGY: An agriculture management paradigm based on ecological rather than solely technological interventions.

ALTERNATE STABLE STATE: See box 3.1.

ANTHROPOCENE: Unofficial name of this epoch in Earth's history.

AQUACULTURE: The cultivation of aquatic organisms in controlled aquatic environments for any commercial, recreational, or public purpose.

ATMOSPHERE: The gaseous volume over a planet's surface where the mass of the planet is large enough to provide the gravitational forces necessary to prevent surface gases from escaping. An atmosphere's composition is a function of many physical and chemical processes and on Earth is also influenced by biological processes such as respiration and photosynthesis.

BACKWARD VERTICAL INTEGRATION: When a business gains control of links in the supply chain connecting the farm to the agribusiness.

BIOCIDES: Chemical agents used to control pests or unwanted species (herbicides, fungicides, and insecticides).

BIOCONTROL AGENT: An organism (the "agent" or "enemy") used to control another organism (e.g., pests, weeds, and plant diseases) through mechanisms such as herbivory, predation, or parasitism.

BIODIVERSITY: The diversity of life on Earth in all its dimensions (e.g., taxonomic, genetic, spatial, temporal, etc.).

BIOGEOCHEMISTRY: The fundamental science concerning biologically driven geochemical processes.

BIOMASS: Within the context of this primer, the total amount, or mass, of organic matter (living and dead); often expressed as the mass per unit of area or volume of habitat or ecosystem and sometimes specifically refers to the mass of organic carbon present (e.g., Megagrams [Mg] of carbon per hectare [ha] of forest).

BIOME: A large-scale, geographically and climatically defined region often containing a biota characteristic to the region. Examples include deserts, savannas, tropical rain forests, coral reefs, continental shelves, and the deep ocean.

BIOPRODUCT: Any biological product produced by farming, whether edible products such as grains, dairy products, or meat; inedible products such as timber, biofuels, cotton, and silk; or drugs such as medicinal herbs, coca (for cocaine), poppy seed pods (for heroin), or marijuana.

BIOSPHERE: The entirety of all space on Earth within which the physical and chemical conditions can support at least one form of life. The biosphere is 3.7 billion years old and consists of a complex web of 8.7 million plant, animal, and microbial species that are linked to one another by the elements they share. Its mass is estimated to be one trillion tons of organic carbon, which cycles between inorganic and organic forms throughout the atmosphere, hydrosphere, and lithosphere via biological, physical, and chemical processes largely driven by energy input from the sun.

BIOTIC INTERACTION: When the biological activity of one individual influences the biological activity of another. Predation, parasitism, competition, and facilitation are common examples of biotic interactions among individuals, populations, and species.

BLUE CARBON: Carbon sequestered (i.e., captured and stored) by oceans and coastal systems.

CAFO: See *concentrated animal feeding.*

CARBON SEQUESTRATION: The process of capturing atmospheric carbon dioxide (CO_2, a greenhouse gas) and storing it long term in plants, oceans, soils, and geologic formations. Carbon sequestration may occur naturally (e.g., by plants, through the process of photosynthesis) or artificially (e.g., capturing CO_2, using pressure to convert it from a gas to a liquid, and injecting the liquid into geologic formations for storage).

CATASTROPHIC CHANGE: See box 3.1.

CLIMATE-SMART AGRICULTURE: See box 3.2

COLLAPSE (ECOLOGICAL OR ECOSYSTEM): See box 3.1.

COMMUNITY-SUPPORTED AGRICULTURE: Directly connecting farmers with consumers by creating contractual agreements in which the farmer provides shares of production to shareholders or subscribers who are usually members of a local community.

CONCENTRATED ANIMAL FEEDING OPERATIONS (CAFOS): High concentrations of confined livestock where animal densities may exceed 100 times that of rangelands.

CULTURAL SERVICES: Ecosystem services defined by peoples' perceptions and beliefs. These may change as culture evolves.

DEVELOPMENT: The process of growth and change that leads to a more mature or advanced state. In the social sciences, development refers to the set of activities undertaken to improve our well-being.

DISTURBANCE (ECOLOGICAL): When an event causes an ecosystem to deviate significantly from its functional norms, such as a catastrophic fire leading to nitrogen loss and mudslides in a terrestrial system or a major oil spill leading to ecosystem collapse in a marine system.

DRIVER: See *stressor*.

ECOLOGICAL BALANCE: Interactions among species and their environment in natural systems generally involve feedbacks that allow populations and biogeochemical processes in ecosystems to fluctuate but stay within upper and lower boundaries such that long-term functioning, or ecological balance, is sustained. Extinction or system collapse is rare.

ECOLOGICAL STABILITY: When environmental factors reach levels significantly above or below long-term norms, such as unprecedented fire, drought, or flood, an ecological system is considered stable if it resists change in the face of extreme conditions or can recover quickly (i.e., within a few generations of the dominant species).

ECOSYSTEM STABILITY: See box 3.1.

ECOREGION: Areas within which ecological factors, such as solar radiation, humidity, temperature, flora, and fauna, are relatively homogeneous.

ECOSERVICES: See ECOSYSTEM SERVICES in box 2.1.

ECOSYSTEM: See box 2.1.

ECOSYSTEM COMPOSITION: See box 2.1.

ECOSYSTEM DISSERVICES: See box 2.1.

ECOSYSTEM DYNAMIC PROPERTIES: See box 2.1.

ECOSYSTEM SERVICES: See box 2.1.

EDAPHIC: Associated soil processes or properties.

ELEMENT: The most chemically simple substance that cannot be broken down or converted to a different substance; there are over 100 elements (e.g., carbon, nitrogen, and phosphorus) on Earth that serve as the constituents of matter.

ENVIRONMENTAL SERVICES: See box 2.1.

ENVIRONMENTAL SUSTAINABILITY: The ability of a system to function in a manner that does not lead to changes in the physical, chemical, or biological conditions that would jeopardize life either at a local or global scale.

FARM LABORERS: See box 1.1.

FARMERS: See box 1.1.

FARMING: See box 1.1.

FARMS: See box 1.1.

FINANCIAL CAPITAL: Assets such as savings, loans, or credit and other monetary instruments that can be used in the exchange, trade, or purchase and sale of goods and services.

FOOD: See box 1.1.

FOOD JUSTICE: The rights of farmers to land and the natural resources they need to pursue their livelihoods.

FOOD SECURITY: The right of all people to robust food supplies.

FOOD SOVEREIGNTY: The rights of communities to choose what they eat and how they farm.

FORWARD VERTICAL INTEGRATION (AGRIBUSINESS): A practice in which an agribusiness invests its assets in control of the supply chain links that lead to the consumer.

GENETICALLY MODIFIED ORGANISM (GMO): Any organism (e.g., plant, animal, or microorganism) whose genome (i.e., DNA) has been modified using genetic engineering to produce the expression of desired traits; genetic modification (or engineering) involves inserting genes from one species into another species (i.e., organisms that would not interbreed with each other in nature; see SPECIES).

GREENHOUSE GAS (GHG): A gas that traps heat in the atmosphere, thereby contributing to climate change and the warming of Earth's surface. The primary anthropogenic (caused by human activities) GHGs are carbon dioxide (CO_2), methane (CH_4), nitrous oxide (N_2O), and fluorinated gases.

HABITAT: The natural environment inhabited by an organism or population of organisms; includes physical/nonliving (e.g., soil, light, and temperature) and biological/living (e.g., food and the presence of predators) factors.

HUMAN CAPITAL: The experience, skills, education, know-how, and other knowledge assets that influence production.

HUMAN DEVELOPMENT: The process by which humans develop to their full potential and are free to make choices and pursue opportunities that allow them to live lives they value. Note that human development is not tied to notions of sustainability.

HUMAN WELL-BEING: Although maximizing human well-being is a key objective of sustainable development, it is poorly defined, but most scholars agree that it concerns happiness and the actualization of human potentials. While the Millennium Ecosystem Assessment did not provide a rigorous definition, it identified five key constituents of human well-being: (1) basic material for a good life, (2) health, (3) freedom of choice and opportunities, (4) security, and (5) good social relations. See figure 4.1.

HYDROSPHERE: All bodies of water on Earth, including oceans, seas, and lakes.

INTERCROPPING: Co-planting two or more species.

INTERGENERATIONAL EQUITY: A concept of fairness and justice that requires that each generation ensure that its development does not jeopardize the rights of future generations to develop.

LAND SHARING: Increasing agricultural production by integrating natural and agricultural systems in a single landscape.

LAND SPARING: Increasing agricultural production without encroaching on natural habitats.

LANDSCAPE: An area containing two or more ecosystems.

LITHOSPHERE: Earth's crust and upper mantle.

MACRONUTRIENTS: See box 4.1.

MALNUTRITION: A blanket term used to describe deficiencies (undernutrition), excesses (overnutrition), or imbalances in a person's energy (i.e., kilocalories) and/or nutrient intake.

MANUFACTURED CAPITAL: The physical objects and infrastructure needed for the production of goods and services such as tools, machinery, computers, buildings, and communication, transportation, and energy infrastructures.

MICRONUTRIENTS: See box 4.1.

NATURAL CAPITAL: Stocks of natural assets such as air, water, soil, geologic formations, noncultivated or nondomesticated plants, animals, and microorganisms (e.g., animal pollinators, soil decomposer microbes, animal and microbial biocontrol agents) and, on the negative side, plant pathogens, livestock diseases, and pests such as herbivorous insects and weeds.

NATURE'S SERVICES: See ECOSYSTEM SERVICES in box 2.1.

NICHE: Definitions vary, but the most common definition is the set of ecological conditions necessary for a species to persist in its habitat for multiple generations.

NICHE CONSTRUCTION: The active modification of one's niche. Most organisms can modify their habitats to improve their likelihood of persisting, such as beavers building dams, birds building nests, bees building hives, or corals building reefs. In chapter 1, we consider niche construction to be the ecological equivalent of *development* as used in the social sciences.

NUTRIENTS: See box 4.1.

OVERNUTRITION: When the diet of an individual provides nutrients in excess of what is needed for normal metabolism, growth, and good health.

REGIME SHIFT: See box 3.1.

PARADIGM: A framework of ideas that are compatible with our understanding of the way the world works.

PEDOSPHERE: The soil surface of Earth; it occurs between at the interface among the atmosphere, lithosphere, and terrestrial biosphere.

PERSISTENCE: See box 3.1.

PERTURBATION (ECOLOGICAL OR ENVIRONMENTAL): See *disturbance (ecological or environmental)*.

PILLARS OF SUSTAINABLE DEVELOPMENT: Human or economic development, to be sustainable, must improve human wellbeing by integrating across three sets of factors (pillars or dimensions): (1) social, (2) economic, and (3) environmental.

PILLARS OF HUMAN DEVELOPMENT: From an economic perspective, there are three key elements (or pillars) of human wellbeing: (1) health, (2) education, and (3) access (to information, water, energy, and other resources, natural or otherwise, including, for farmers, land, seed, livestock, fertilizers and biocides, and other forms of capital).

PRESSURE (ECOLOGICAL OR ENVIRONMENTAL): See *stressor*.

PROVISIONING SERVICES: The goods ecosystems provide to humans.

PSEUDOCEREALS: Cereals that do not come from grasses (e.g., amaranth, buckwheat, spelt, teff).

PULSES: Edible seeds of plants in the legume family (e.g., peas, lentils, black-eyed peas, chickpeas/garbanzo beans, kidney beans).

RANGELAND: Terrestrial systems that lack or have too few trees to be called forests. Includes a variety of vegetation, such as deserts, grasslands, savannas (most often a mix of grasses and trees), shrublands, tundra, and wetlands.

REGULATING SERVICES: Ecosystem characteristics that reduce variability or otherwise promote system stability.

RESILIENCE: In ecology, time it takes for a system to return to the state it had before the perturbation; short return times mean higher resilience. In sustainability science, ability to withstand environmental shocks. See box 3.1 for related terms.

RESISTANCE: See box 3.1.

SOCIAL CAPITAL: The trust; connections through friends, families, and colleagues; and membership in groups with shared values, such as religious organizations, that can affect wealth.

SOCIAL COHESION: A complex concept that weaves together ideas about equality and social capital.

SPECIES: A biological classification of organisms that can interbreed with each other to produce fertile offspring.

STABILITY: See box 3.1.

STABLE STATES: See box 3.1.

STRESSOR (ECOLOGICAL OR ENVIRONMENTAL): A chronic or persistent environmental change that can lead to a catastrophic shift or ecosystem collapse once a threshold or tipping point is crossed.

SUBSISTENCE FARMING/AGRICULTURE: A farming system in which all (or nearly all) the food grown and/or livestock raised is used to feed and provide for the farmer's family, thereby leaving little to no surplus to sell or trade.

SUCCESSION (ECOLOGICAL): Return of an ecosystem to its pre-disturbance state.

SUPPLY CHAIN: A system, and its associated processes, to produce, transport, and distribute a good to consumers.

SUPPORTING SERVICES: Ecosystem services that indirectly benefit humans by making other goods and services possible.

SUSTAINABLE DEVELOPMENT: Environmentally sustainable management that improves human well-being while balancing inputs with outputs, ensuring system resilience, and guaranteeing that future generations will be able to do the same.

SUSTAINABLE HUMAN DEVELOPMENT: Human development that includes notions of sustainability in its objectives.

SUSTAINABLE INTENSIFICATION: A process by which agricultural yields are increased without adverse environmental impacts and without the further conversion of remaining natural habitats to agriculture.

SYSTEM: A network of interacting entities, such as interacting people in a social system; interacting plants, animals, and microorganisms in an ecosystem; or interacting components in an engineered system.

SYSTEM DYNAMICS: See box 3.1.

SYSTEM STATE: See box 3.1.

TAXA: Plural of *taxon*; units of organisms ranked hierarchically as a means of classification (e.g., domain, kingdom, phylum, class, order, family, genus, and species).

THRESHOLDS AND TIPPING POINTS: See box 3.1.

TRIPLE BOTTOM LINE: An accounting paradigm in which the net value of an enterprise is tallied using a ledger that includes natural, economic, and social costs and benefits.

2030 AGENDA FOR SUSTAINABLE DEVELOPMENT: Adopted by the United Nations (UN) in 2015, the agenda is a plan to promote peace and prosperity for all while protecting Earth and its resources now and into the future; the seventeen Sustainable Development Goals serve as the roadmap to the sustainable and equitable future envisioned by the UN.

UNDERNUTRITION: When the diet of an individual does not supply nutrients sufficient to maintain normal metabolic processes, growth, and good health.

UNITED NATIONS CONVENTION TO COMBAT DESERTIFICATION: Program that focuses on the complex social and economic factors that drive desertification in dryland rather than focusing on yields.

UNITED NATIONS REDUCING EMISSIONS FROM DEFORESTATION AND FOREST DEGRADATION IN DEVELOPING COUNTRIES (REDD+): A program intended to reduce deforestation while supporting sustainable development.

VARIABILITY: Range of values, usually bounded by upper and lower limits, that state variables can have over time.

VERTICAL INTEGRATION (AGRIBUSINESS): A practice in which an agribusiness invests its assets in acquiring or controlling the supply chain. *Backward vertical integration* involves gaining control of links in the supply chain that lead to the farmer; *forward vertical integration* involves gaining control of the supply chain links that lead to the consumer.

VITAMINS: See box 4.1.

YIELD GAPS: The difference between the actual yield of a unit of farmland and its potential yield if it were managed under optimal conditions of nutrient supply, water, and protection against pathogens and pests.

NOTES

1. SUSTAINABLE DEVELOPMENT: A NEW CENTURY, A NEW PARADIGM

1. Note that a metric ton (or tonne), a U.S. ton, and an imperial ton are close enough that we won't worry about conversions between U.S. and metric values as we do with grams and ounces or kilograms and pounds.
2. Yinon M. Bar-On, Rob Phillips, and Ron Milo, "The Biomass Distribution on Earth," *Proceedings of the National Academy of Sciences* (2018).
3. Food and Agriculture Organization of the United Nations, FAO-STAT Database, http://www.fao.org/faostat/en/#data/EK.
4. United Nations Department of Economic and Social Affairs, Population Division, *World Population Prospects: The 2015 Revision: Key Findings and Advance Tables*, Working Paper no. ESA/P/WP/248 (New York: UN DESA, 2017).
5. World Commission on Environment and Development, *Our Common Future* (Oxford: Oxford University Press, 1987).
6. Food and Agriculture Organization of the United Nations, "SAFA: Sustainability Assessment of Food and Agriculture Systems: Guidelines," Rome: Natural Resources Management and Environment Department, 2012.
7. Junguo Liu, Kun Ma, Philippe Ciais, and Stephen Polasky, "Reducing Human Nitrogen Use for Food Production," *Scientific Reports* 6 (July 2016): 30104; W. H. Schlesinger and E. S. Bernhardt,

Biogeochemistry: An Analysis of Global Change, 3rd ed. (Cambridge, MA: Academic Press, 2012); P. M. Vitousek et al., "Human Alteration of the Global Nitrogen Cycle: Sources and Consequences," *Ecological Applications* 7 (1997): 737–50.

8. J. Rockström et al., "A Safe Operating Space for Humanity," *Nature* 461 (2009): 472–75.

9. W. Steffen et al., "Planetary Boundaries: Guiding Human Development on a Changing Planet," *Science* 347, no. 6223 (2015).

10. Peter Rosset, "Re-Thinking Agrarian Reform, Land, and Territory in la Via Campesina," *Journal of Peasant Studies* 40 (2013): 721–75.

11. A. H. Alkon and T. M. Mares, "Food Sovereignty in US Food Movements: Radical Visions and Neoliberal Constraints," *Agriculture and Human Values* 29 (2012).

12. D. Tilman, K. G. Cassman, P. A. Matson, R. Naylor, and S. Polasky, "Agricultural Sustainability and Intensive Production Practices," *Nature* 418, no. 6898 (August 2002): 671–77.

13. Ruth DeFries, Jessica Fanzo, Roseline Remans, et al., "Metrics for Land-Scarce Agriculture," *Science* 349, no. 6245 (2015): 238–40.

14. Millennium Ecosystem Assessment, *Ecosystems and Human Well-Being: Synthesis* (Washington, DC: Island, 2005); F. Isbell et al., "Benefits of Increasing Plant Diversity in Sustainable Agroecosystems," *Journal of Ecology* 105 (2017): 871–79.

15. Walter Willett, Johan Rockström, Brent Loken, et al., "Food in the Anthropocene: The Eat–Lancet Commission on Healthy Diets from Sustainable Food Systems," *The Lancet* 393, no. 10170 (2019): 447–92.

16. Prabhu L. Pingali, "Green Revolution: Impacts, Limits, and the Path Ahead," *Proceedings of the National Academy of Sciences of the United States of America* 109, no. 31 (2012): 12302–8.

17. Ivette Perfecto, John Vandermeer, and Angus Wright, *Nature's Matrix: Linking Agriculture, Conservation, and Food Sovereignty* (New York: Routledge, 2009).

18. Brenda B. Lin, "Resilience in Agriculture Through Crop Diversification: Adaptive Management for Environmental Change," *BioScience* 61, no. 3 (2011): 183–93.

19. Mark A. Boudreau, "Diseases in Intercropping Systems," *Annual Review of Phytopathology* 51 (2013): 499–519.

20. Ruth DeFries, Jessica Fanzo, Roseline Remans, et al., "Metrics for Land-Scarce Agriculture," *Science* 349, no. 6245 (2015): 238–40.

21. Anamarija Frankic and Carl Hershner, "Sustainable Aquaculture: Developing the Promise of Aquaculture," *Aquaculture International* 11, no. 6 (November 2003): 517–30; Sara Tjossem, *Fostering Internationalism Through Marine Science: The Journey with Pisces* (New York: Springer, 2016).

2. NATURE AND NATURE'S GOODS AND SERVICES

1. United States Department of Labor Bureau of Labor Statistics, *Employment by Major Industry Sector* (Washington, DC: U.S. Department of Labor, 2018).

2. United Nations Conference on Trade and Development, "Beyond Austerity: Towards a Global New Deal," Trade and Development Report, Geneva, 2017.

3. United States Environmental Protection Agency (EPA), "Ecoregions," 2018; David M. Olson, Eric Dinerstein, Eric D. Wikramanayake, et al., "Terrestrial Ecoregions of the World: A New Map of Life on Earth," *Bioscience* 51, no. 11 (2001): 933.

4. Pushpam Kumar, *The Economics of Ecosystems and Biodiversity: Ecological and Economic Foundations* (London: Routledge, 2012); UK National Ecosystem Assessment (UK-NEA), *The UK National Ecosystem Assessment Technical Report* (Cambridge: UN Environmental Programme— World Conservation Monitoring Centre [UNEP-WCMC], 2011); R. Haines-Young and M. B. Potschin, *Common International Classification of Ecosystem Services (CICES) V5.1 and Guidance on the Application of the Revised Structure* (Nottingham: United Nations Statistical Division, 2017).

5. Millennium Ecosystem Assessment, *Ecosystems and Human Well-Being: Synthesis* (Washington, DC: Island, 2005).

6. For a more thorough explanation of ecosystem services, see G. C. Daily, ed., *Nature's Services* (Washington, DC: Island, 1997).

7. Rodney J. Keenan, Gregory A. Reams, Frédéric Achard, et al., "Dynamics of Global Forest Area: Results from the Fao Global

Forest Resources Assessment 2015," *Forest Ecology and Management* 352 (2015): 9–20.

8. K. MacDicken, Ö. Jonsson, L. Piña, et al., *Global Forest Resources Assessment 2015: How Are the World's Forests Changing?* (FAO, 2016).

9. United States Department of Agriculture and United States Environmental Protection Agency (USDA and USEPA), "Unified National Strategy for Animal Feeding Operations," USDA and USEPA, Washington, DC, 1999.

10. United Nations Convention to Combat Desertification (UNCCD), "Global Land Outlook," Bonn, 2017.

11. US Environmental Protection Agency (USEPA), "Ag 101," 2015.

12. United States Department of Agriculture Foreign Agricultural Service (USDA-FAS), *World Agricultural Production*, WAP 5-19 (USDA, 2019).

13. B. Worm, E. B. Barbier, N. Beaumont, et al., "Impacts of Biodiversity Loss on Ocean Ecosystem Services," *Science* 314, no. 5800 (November 2006): 787–90.

14. Food and Agriculture Organization (FAO), *The State of World Fisheries and Aquaculture 2016: Contributing to Food Security and Nutrition for All* (Rome: UN Food and Agriculture Organization, 2016).

15. K. E. Charlton, J. Russell, E. Gorman, et al., "Fish, Food Security, and Health in Pacific Island Countries and Territories: A Systematic Literature Review," *BMC Public Health* 16 (2016): 285.

16. Ragnar Arnason, *The Economics of Ocean Ranching: Experiences, Outlook, and Theory* (Food & Agriculture Org., 2001).

17. Ashifa Kassam, "Thousands of Atlantic Salmon Escape from Fish Farm Into the Pacific," *The Guardian*, August 23, 2017.

18. Oren Shelef, Peter J. Weisberg, and Frederick D. Provenza, "The Value of Native Plants and Local Production in an Era of Global Agriculture," *Frontiers in Plant Science* 8 (2017): 2069.

19. FAO, *The State of World Fisheries and Aquaculture 2016.*

20. R. Costanza, R. D'Arge, R. de Groot, et al., "The Value of the World's Ecosystem Services and Natural Capital," *Nature* 387 (1997): 253–60; R. Costanza, R. de Groot, P. Sutton, et al., "Changes in the Global Value of Ecosystem Services," *Global Environmental Change* 26 (2014): 152–58.

21. J. A. Foley, R. DeFries, G. P. Asner, et al., "Global Consequences of Land Use," *Science* 309 (2005): 570–74.

3. SUSTAINABLE DEVELOPMENT AND FOOD PRODUCTION

1. Thomas H. Kunz, Elizabeth Braun de Torrez, Dana Bauer, et al., "Ecosystem Services Provided by Bats," *Annals of the New York Academy of Sciences* 1223, no. 1 (2011): 1–38.

2. Ransom A. Myers, Jeffrey A. Hutchings, and Nicholas J. Barrowman, "Why Do Fish Stocks Collapse? The Example of Cod in Atlantic Canada," *Ecological Applications* 7, no. 1 (1997): 91–106.

3. Uriel Safriel and Zafar Adeel, "Development Paths of Drylands: Thresholds and Sustainability," *Sustainability Science* 3, no. 1 (2008): 117–23.

4. Rhys E. Green, Stephen J. Cornell, Jörn P. W. Scharlemann, and Andrew Balmford, "Farming and the Fate of Wild Nature," *Science* 307, no. 5709 (2005): 550–55.

5. Joern Fischer, David J. Abson, Van Butsic, et al., "Land Sparing Versus Land Sharing: Moving Forward," *Conservation Letters* 7, no. 3 (2014): 149–57; Elena M. Bennett, "Changing the Agriculture and Environment Conversation," *Nature Ecology and Evolution* 1, no. 1 (2017): 1–2; Ole Mertz and Charlotte Filt Mertens, "Land Sparing and Land Sharing Policies in Developing Countries—Drivers and Linkages to Scientific Debates," *World Development* 98 (2017): 523–35.

6. United Nations Food and Agriculture Organization (FAO), "'Climate-Smart' Agriculture: Policies, Practices, and Financing for Food Security, Adaptation, and Mitigation," Rome, 2010.

7. United Nations Food and Agriculture Organization (FAO), "Climate Smart Agriculture: Building Resilience to Climate Change," ed. L. Lipper, N. McCarthy, D. Zilberman, et al., Rome, 2018; Leslie Lipper, Philip Thornton, Bruce M. Campbell, et al., "Climate-Smart Agriculture for Food Security," *Nature Climate Change* 4, no. 12 (2014): 1068.

8. Andrés Castañeda, Dung Doan, David Newhouse, et al., *Who Are the Poor in the Developing World?* (Washington, DC: World Bank, 2016);

United Nations Food and Agriculture Organization (FAO), *The State of Food and Agriculture: Innovation in Family Farming* (Rome: FAO, 2014).

9. Olivier De Schutter, "Report of the Special Rapporteur on the Right to Food." United Nations Food and Agriculture Organization (FAO), Rome, 2014.

10. Gill Seyfang, "Shopping for Sustainability: Can Sustainable Consumption Promote Ecological Citizenship?" *Environmental Politics* 14, no. 2 (2005): 290–306; Lucia Reisch, Ulrike Eberle, and Sylvia Lorek, "Sustainable Food Consumption: An Overview of Contemporary Issues and Policies," *Sustainability: Science, Practice and Policy* 9, no. 2 (2013): 7–25; Jennifer L. Jacquet and Daniel Pauly, "The Rise of Seafood Awareness Campaigns in an Era of Collapsing Fisheries," *Marine Policy* 31, no. 3 (2007): 308–13.

11. Greenpeace, "About Greenpeace," 2019, https://www.greenpeace.org /usa/sustainable-agriculture/; Bioversity International, "About Us," 2019, https://www.bioversityinternational.org/; Center for Food Safety, "About Us," 2019, http://www.centerforfoodsafety.org/; Good Planet Foundation, "About Us," https://www.goodplanet.org/en/; Food and Water Watch, "About," 2019. https://www.foodandwaterwatch.org/.

12. Jeetendra P. Aryal and Stein T. Holden, "Livestock and Land Share Contracts in a Hindu Society," *Agricultural Economics* 43, no. 5 (2012): 593–606.

13. John Emmeus Davis, *The Community Land Trust Reader* (Lincoln Institute of Land Policy, 2010).

14. Mark Paul, "Community-Supported Agriculture in the United States: Social, Ecological, and Economic Benefits to Farming," *Journal of Agrarian Change* 19, no. 1 (2019): 162–80.

15. Kai-Ingo Voigt, Oana Buliga, and Kathrin Michl, "Globalizing Coffee Culture: The Case of Starbucks," in *Business Model Pioneers: How Innovators Successfully Implement New Business Models*, ed. Kai-Ingo Voigt, Oana Buliga and Kathrin Michl, 41–53 (Cham: Springer International, 2017); Starbucks Company Profile, 2019, https://www .starbucks.com/about-us/company-information/starbucks-company -profile.

16. World Business Council for Sustainable Development, "About Us," 2019, https://www.wbcsd.org/Overview/About-us.

17. Stephen R. Gliessman, *Agroecology: The Ecology of Sustainable Food Systems* (CRC, 2014).

18. E. Holt Giménez, *Campesino a Campesino: Voices from Latin America's Farmer-to-Farmer Movement for Sustainable Agriculture* (Food First, 2006); La Via Campesina, "Who Are We," 2019, https://viacampesina.org/en/who-are-we/; Hannah Wittman, "Reframing Agrarian Citizenship: Land, Life, and Power in Brazil," *Journal of Rural Studies* 25, no. 1 (2009): 120–30; National Young Farmer's Coalition, "About," 2019, https://www.youngfarmers.org/about/; Association for India's Development, "About AID," 2019, https://aidindia.org/about-aid/.

19. Jennifer Clapp, *Food*, 1st ed. (Cambridge: Polity, 2011).

20. World Bank, "World Development Report. Agriculture for Development," 2008.

21. Madeleine Fairbairn. "'Like Gold with Yield': Evolving Intersections Between Farmland and Finance," *Journal of Peasant Studies* 41, no. 5 (2014): 777–95.

22. J. L. Combes, T. Kinda, R. Ouedraogo, and P. Plane, "Does It Pour When It Rains? Capital Flows and Economic Growth in Developing Countries," CERDI, Études et Documents 2, 2017.

23. Clapp, *Food*.

24. United Nations Conference on Trade and Development (UNCTAD), "World Investment Report: Transnational Corporations, Agricultural Production, and Development," UNCTAD, 2009.

25. E. Andrews, "Why You Should Be Skeptical of Walmart's Cheap Organic Food," *Grist*, 2014, https://grist.org/food/why-you-should-be-skeptical-of-walmarts-cheap-organic-food/.

26. Clapp, *Food*.

27. World Trade Organization, "The Uruguay Round," 2019, https://www.wto.org/english/thewto_e/whatis_e/tif_e/fact5_e.htm.

28. Tim Josling and Stefan Tangermann, "Implementation of the WTO Agreement on Agriculture and Developments for the Next Round of Negotiations," *European Review of Agricultural Economics* 26, no. 3 (1999): 371–88.

29. S. Murphy, B. Lillison, and M. B. Lake, "WTO Agreement on Agriculture: A Decade of Dumping," Institute for Agriculture and Trade Policy, Minneapolis, MN, 2005.

30. Food and Agriculture Organization of the United Nations, "World Trade Organization (WTO) Agreement on Agriculture: Export Competition After the Nairobi Ministerial Conference," FAO, 2017.

31. Andy Gutierrez, "Codifying the Past, Erasing the Future: NAFTA and the Zapatista Uprising of 1994," *Hastings Environmental Law Journal* 4 (1998): 143–62.

32. Office of the United States Trade Representative, "United States–Mexico-Canada 2019 Agreement," https://ustr.gov/usmca.

4. FOOD, FARMING, AND HUMAN WELL-BEING

1. United Nations Development Programme (UNDP), *Human Development Report 1997* (New York: Oxford University Press, 1997).

2. Amartya Sen, *Development as Freedom* (New York: Knopf, 1999).

3. Richard M. Ryan and Edward L. Deci, "On Happiness and Human Potentials: A Review of Research on Hedonic and Eudaimonic Well-Being," *Annual Review of Psychology* (2001).

4. Rachel Dodge, Annette P. Daly, Jan Huyton, and Lalage D. Sanders, "The Challenge of Defining Wellbeing," *International Journal of Wellbeing* 2, no. 3 (2012).

5. Joseph Alcamo, *Ecosystems and Human Well-Being: A Framework for Assessment* (Washington, DC: Island, 2003).

6. UNFAO, WHO, and United Nations University, "Human Energy Requirements. Report. A Joint FAO-WHO-UNU Expert Consultation," Food and Agriculture Organization of the United Nations, Rome, 2004, http://www.fao.org/3/y5686e/y5686e00.htm.

7. United Nations World Health Organization (WHO), "Healthy Diet," 2018, https://www.who.int/news-room/fact-sheets/detail/healthy-diet.

8. United Nations Food and Agriculture Organization (FAO), *What Is Agrobiodiversity? Building on Gender, Agrobiodiversity, and Local Knowledge* (Rome: UNFAO, 2004).

9. World Health Organization (WHO), "The Double Burden of Malnutrition: Policy Brief," WHO/NMH/NHD/17.3, Department of Nutrition for Health and Development, WHO, 2017.

10. United Nations Children's Fund (UNICEF), World Health Organization (WHO), and World Bank Group (WBG), *Levels and Trends in Child Malnutrition: Key Findings of the 2018 Edition of the Joint Child Malnutrition Estimates* (New York: United Nations, 2018).

11. NCD Risk Factor Collaboration (NCD-RisC), "Worldwide Trends in Body-Mass Index, Underweight, Overweight, and Obesity from 1975 to 2016: A Pooled Analysis of 2,416 Population-Based Measurement Studies in 128.9 Million Children, Adolescents, and Adults," *The Lancet* 390 (2017): 2627–42.

12. Gregory A. Roth, Degu Abate, Kalkidan Hassen Abate, et al., "Global, Regional, and National Age-Sex-Specific Mortality for 282 Causes of Death in 195 Countries and Territories, 1980–2017: A Systematic Analysis for the Global Burden of Disease Study 2017," *The Lancet* 392, no. 10159 (2018): 1736–88.

13. The U.S. Burden of Disease Collaborators, "The State of U.S. Health, 1990–2016: Burden of Diseases, Injuries, and Risk Factors among U.S. States," *JAMA* 319, no. 14 (2018): 1444–72.

14. Center for Disease Control (CDC), "Food Safety: Foodborne Illnesses and Germs," 2018, https://www.cdc.gov/foodsafety/foodborne-germs .html.

15. UN General Assembly, *Transforming Our World: The 2030 Agenda for Sustainable Development, 21 October 2015*, A/RES/70/1, 2015, https:// www.refworld.org/docid/57b6e3e44.html.

16. United Nations Food and Agriculture Organization (FAO), *Edible Insects: Future Prospects for Food and Feed Security* (Rome: FAO, 2013).

17. United Nations Food and Agriculture Organization (FAO), *The State of Food Insecurity in the World 2001* (Rome: FAO, 2001).

18. United States Department of Agriculture Economic Research Service (USDA ERS), "Definitions: Food Access," USDA, 2019, https://www.ers.usda.gov/data-products/food-access-research-atlas /documentation/.

19. Hannah Wittman, Michael Jahi Chappell, David James Abson, et al., "A Social-Ecological Perspective on Harmonizing Food Security and Biodiversity Conservation," *Regional Environmental Change* 17, no. 5 (2017): 1291–301.

5. BRINGING IT ALL TOGETHER: THE SUSTAINABLE FUTURE BEFORE US

1. E. Bennett Carys, E., Richard Thomas, Mark Williams, et al., "The Broiler Chicken as a Signal of a Human Reconfigured Biosphere," *Royal Society Open Science* 5, no. 12: 180325.

2. John Asafu-Adjaye, Linus Blomqvist, Stewart Brand, et al., "An Eco-modernist Manifesto," 2015, http://www.ecomodernism.org/.

3. Walter Willett, Johan Rockström, Brent Loken, et al., "Food in the Anthropocene: The Eat-Lancet Commission on Healthy Diets from Sustainable Food Systems," *The Lancet* 393, no. 10170 (2019): 447–92.

4. United Nations Food and Agriculture Organization (FAO), *The State of World Fisheries and Aquaculture 2018—Meeting the Sustainable Development Goals* (Rome: FAO, 2018), CC BY-NC-SA 3.0 IGO.

5. P. Behrens, J. C. Kiefte-de Jong, T. Bosker, et al., "Evaluating Environmental Impacts of Dietary Recommendations," *Proceedings of the National Academies of Science* 114, no. 51 (December 19, 2017): 13412–17.

6. J. A. Foley, N. Ramankutty, K. A. Brauman, et al., "Solutions for a Cultivated Planet," *Nature* 478 (2011): 337–42.

7. Karl-Heinz Erb, Christian Lauk, Thomas Kastner, et al., "Exploring the Biophysical Option Space for Feeding the World Without Deforestation," *Nature Communications* 7 (2016): 11382.

8. International Panel of Experts on Sustainable Food Systems (IPES-Food), *Breaking Away from Industrial Food and Farming Systems: Seven Case Studies of Agroecological Transitions*, IPES-Food, Case Studies 02, 2018; IPES-Food, *From Uniformity to Diversity: A Paradigm Shift from Industrial Agriculture to Diversified Agroecological Systems*, IPES-Food, Report 2, 2016.

9. Christophe Béné, Peter Oosterveer, Lea Lamotte, et al., "When Food Systems Meet Sustainability—Current Narratives and Implications for Actions," *World Development* 113 (2019): 116–30.

11. UN General Assembly, *Transforming Our World: The 2030 Agenda for Sustainable Development*, October 21, 2015, A/RES/70/1, https://sustainabledevelopment.un.org/post2015/transformingourworld; Willett et al., "Food in the Anthropocene."

12. United Nations Development Programme (UNDP), "Sustainable Development Goals," https://www.undp.org/content/undp/en/home

/sustainable-development-goals.html; United Nations Department of Economic and Social Affairs (UNDESA), "SDG Indicators," UNDESA, Statistics Division, 2019, https://unstats.un.org/sdgs /indicators/indicators-list/.

13. David Le Blanc, "Towards Integration at Last? The Sustainable Development Goals as a Network of Targets," *Sustainable Development* 23, no. 3 (2015): 176–87.

14. Sarah Murray, "Fixing Food 2018: Best Practices Towards the Sustainable Development Goals," ed. Martin Koehring, Economist Intelligence Unit, 2018, https://foodsustainability.eiu.com/whitepaper -2018/.

16. UNDESA, *Transforming Our World*; D. Collste, M. Pedercini, and S. E. Cornell, "Policy Coherence to Achieve the SDGs: Using Integrated Simulation Models to Assess Effective Policies," *Sustainability Science* 12 (2017): 921–31.

17. In April 2019, shortly before the launch of the EAT-Lancet Commission's report, the World Health Organization withdrew its endorsement of the report. The WHO found the report's strong emphasis on severely reducing livestock production and meat consumption as policy recommendations that would engender considerable hardship for many developing nations. It found the universal diet used by the report in its analyses to be scientifically untenable and critically deficient in several key nutrients. The WHO also found that the commission's funding (in particular, the Stordalen Foundation, the Wellcome Trust, Food Reform for Sustainability and Health, and the World Business Council for Sustainable Development) and its membership (though some of its members were, in fact, officials from the WHO and FAO) compromised the alleged neutrality of the report.

18. United Nations Development Group, "A Million Voices: The World We Want," 2013.

19. John P. Reganold and Jonathan M. Wachter, "Organic Agriculture in the Twenty-First Century," *Nature Plants* 2, no. 2 (2016): 15221.

20. Verena Seufert and Navin Ramankutty, "Many Shades of Gray: The Context-Dependent Performance of Organic Agriculture," *Science Advances* 3, no. 3 (2017): e1602638; Reganold and Wachter, "Organic Agriculture in the Twenty-First Century"; Lauren C. Ponisio,

Leithen K. M'Gonigle, Kevi C. Mace, et al., "Diversification Practices Reduce Organic to Conventional Yield Gap," *Proceedings of the Royal Society B: Biological Sciences* 282, no. 1799 (2015): 20141396; Frank Eyhorn, Adrian Muller, John P. Reganold, et al., "Sustainability in Global Agriculture Driven by Organic Farming," *Nature Sustainability* 2, no. 4 (2019): 253–55.

21. H. Godfray and J. Charles, "The Debate Over Sustainable Intensification," *Food Security* 7, no. 2 (2015): 199–208.

22. Jules Pretty et al., "Intensification for Redesigned and Sustainable Agricultural Systems," *Science* 362, no. 6417 (2018): eaav0294.

23. Laura Diaz Anadon, Gabriel Chan, Alicia G. Harley, et al., "Making Technological Innovation Work for Sustainable Development." *Proceedings of the National Academy of Sciences* 113, no. 35 (2016): 9682.

24. Mario Herrero, Philip K. Thornton, Brendan Power, et al., "Farming and the Geography of Nutrient Production for Human Use: A Transdisciplinary Analysis," *Lancet Planetary Health* 1, no. 1 (2017): e33–e42.

INDEX

access to water, sanitation, and
hygiene (WASH), 130–31
ADM. *See* Archer Daniels
Midland
AFOs. *See* animal feeding operations
Agreement on Agriculture,
Uruguay Round (AoA), 99–100
agribusiness: vertical integration
in, 92. *See also* industrial
farming
agricultural development: under
Agreement on Agriculture,
Uruguay Round, 99–100;
during Anthropocene epoch,
137–41; Archer Daniels
Midland, 97; centralized
market controls in, 97–98;
ecomodernism and, 139–40;
food production and, 95–101;
food security and, 96; future
directions of, 101; under
General Agreement on Tariffs
and Trade, 99; during Industrial
Revolution, 139; international

aid and, 96–97; international
investments in, 96–97; maize
production and, 100; under
North American Free Trade
Agreement, 100; soybean
production and, 97; trade
agreements and policies for,
98–101; undernourishment and,
140; in United States, 97; under
United States–Mexico-Canada
Agreement, 100; World Trade
Organization governance of, 99
agriculture: climate-smart, 84–86;
community-supported, 92;
definition of, 6; economic
dimensions of, 23–27;
environmental sustainability of,
xi; Green Revolution, 31–33,
163–64; labor in, 26; Mayan, 77;
practices in, 6; productivity of,
75–76; social dimension of, 26–27;
sustainable, 33–35, 146–48, 165–66;
sustainable farming network,
28–29; unsustainable, 31–33

and, 14–15, 19. *See also*
biochemistry
biomass: carbon as element of, 3; of
species, measurements for, 3–4
biomes: ecosystems as element of,
2, 40; subsystems of, 2
biomolecules, 19; food as source of,
20
bioproducts, 6–7, 9, 28–29, 35,
47, 75
biosphere: agroecosystems as
dominant element of, viii;
biome subsystems, 2; biotic
interactions, 2; carbon in, 21;
Homo sapiens in, 1–5; species
success in, factors for, 2–5
Biosphere 2, ix–x
Biospherians, ix–x
biotic interaction, 2
blue carbon, 56
Blue Revolution, 59
boundary transgression, 26
Brazilian Patanal, *61,* 64–65
Brundtland, Gro, 11
Brundtland Report, 11–13; farming
in, 14; intergenerational equity,
12
businesses: participation in social
sustainability by, 91–93; supply
chains, 91
bycatch, 62

CAFO. *See* concentrated animal
feeding
calories, 114–15
Canada, 100

capital: financial, 23, *24;* human,
24, *24;* manufactured, 23–24, *24;*
natural, *24,* 24–26, 75; social, 24,
24, 93
capture fisheries. *See* fisheries
carbon (C), 15, *16–17;* in biosphere,
21; blue, 56; as element of
biomass, 3
carbon dioxide (CO_2), ix;
anthropogenic emissions, 57;
carbon sequestration, 56; in
input/output balance, 74
carbon sequestration, 56
carnivores, 63
catastrophic change, 79
Center for Food Safety, 89
chemical contamination, of food,
129–31
China, intercropping in, 34
CICES. *See* Common
International Classification of
Ecosystem Services
climate-smart agriculture, 84–86;
flexibility of, 86; practices
of, 85
CO_2. *See* carbon dioxide
coastal systems: blue carbon in, 56;
carbon sequestration, 56; as
ecosystem, 55–56
Common International
Classification of Ecosystem
Services (CICES), 43
community-supported agriculture,
92
concentrated animal feeding
operations (CAFO), 50–51, 66

unsustainable agriculture, 31–33
U.S. *See* United States
USMCA. *See* United
  States–Mexico-Canada
  Agreement

vertical integration: in agribusiness,
  92; backward, 92; forward, 92
Via Campesina, la, 28
vitamins, 19, 120–21

WASH. *See* access to water,
  sanitation, and hygiene
waste generation, 29
wasting, 115
water sources: for *Homo sapiens,*
  appropriation of rainwater by, 4;
  quality of, regulation
  mechanisms for, 59
WBCSD. *See* World Business
  Council for Sustainable
  Development
well-being, of humans: definitions
  of, 105; food and, 108–21; good
  life and, 113–14; human
  development and, 104–7; in
  Millennium Ecosystem
  Assessment framework, 105,
  106; UN Sustainable
  Development Goals, 107

wetlands: Brazilian Patanal, *61,*
  64–65; as ecosystem, 42
WFP. *See* World Food Program
WHO. *See* World Health
  Organization
World Business Council for
  Sustainable Development
  (WBCSD), 93
World Commission on
  Environment and
  Development, 11
World Food Program (WFP), 27
World Forum of Fisher Peoples,
  28
World Health Organization
  (WHO), 122–25, 195n17; on food
  inequity, 141
World March of Women, 28
World Trade Organization
  (WTO), 99
World Wildlife Fund (WWF),
  40
WTO. *See* World Trade
  Organization
WWF. *See* World Wildlife Fund

years lost to disability (YLD), 126,
  127
yield gaps, 143, 160
YLD. *See* years lost to disability